John Peter Wade

**A Paper on the Prevention and Treatment of the Disorders**

of the Seamen and Soldiers in Bengal: presented to the Honourable Court of

East-India Directors, in the year 1791

John Peter Wade

**A Paper on the Prevention and Treatment of the Disorders**
*of the Seamen and Soldiers in Bengal: presented to the Honourable Court of East-India Directors, in the year 1791*

ISBN/EAN: 9783337309213

Printed in Europe, USA, Canada, Australia, Japan

Cover: Foto ©berggeist007 / pixelio.de

More available books at **www.hansebooks.com**

A

# P A P E R

ON THE

## PREVENTION AND TREATMENT

OF THE

## DISORDERS of SEAMEN and SOLDIERS

IN

## B E N G A L.

PRESENTED TO THE HONOURABLE COURT OF EAST-
INDIA DIRECTORS, IN THE YEAR 1791.

By JOHN PETER WADE, M. D.

L O N D O N:

PRINTED FOR J. MURRAY, N° 32, FLEET STREET.

M.DCC.XCIII.

# DEDICATION.

TO THE HONOURABLE

# COURT of DIRECTORS

OF THE

## *EAST-INDIA COMPANY*;

FRANCIS BARING, Esq. Chairman.
JOHN SMITH BURGES, Esq. Deputy
Chairman.

WILLIAM BENSLEY, Esq.
JACOB BOSANQUET, Esq.
THOMAS CHEAP, Esq.
LIONEL DARELL, Esq.
WILLIAM DEVAYNES, Esq.
WILLIAM ELPHINSTONE, Esq.
WALTER EWER, Esq.
THOMAS FITZHUGH, Esq.
JOHN HUNTER, Esq.
HUGH INGLIS, Esq.
PAUL LE MESURIER, Esq.
Sir STEPHEN LUSHINGTON, Bart.
JOHN MANSHIP, Esq.
THOS THEOPH. METCALF, Esq.
WILLIAM MONEY, Esq.
THOMAS PATTLE, Esq.
JOHN ROBERTS, Esq.
NATHANIEL SMITH, Esq.

ROBERT THORNTON, Esq.
JOHN TOWNSON, Esq.
JOHN TRAVERS, Esq.
STEPHEN WILLIAMS, Esq.


HONOURABLE SIRS,

PERMIT me to intreat your favourable acceptance of the following pages, and to exprefs an humble hope, that they may be an additional, though inconfiderable, proof of the attention of the Medical Department to the interefts of the fervice.

I have the honour to be,
With great refpect,

HONOURABLE SIRS,
Your moft devoted
And moft obedient humble fervant,

JOHN PETER WADE.

Sept. 1792.

# CONTENTS.

# CONTENTS.

# E R R A T A:

Page 6. line 6.  *After* decks,  *add*—Little deference, however, fhould be paid to the teftimony of a fingle inftance, although it afforded even a more pofitive evidence than the prefent againft the exiftence of contagion in fhip fever.  It does not feem prudent or poffible to fpeak with much affurance on either fide of the queftion. The fame facts will fuggeft oppofite conclufions to the reafoning faculties of different perfons; and circumftances, which failed to afford the writer a conviction of the prefence of contagion, would probably furnifh other practitioners with the ftrongeft poffible proofs of the reality of its exiftence.

Page 45. l. 9. Fevers, *add as a note*—The following obfervations on Fevers, with fome alterations, were prefented to the Government of Bengal, in the year 1788.

ON THE

## PREVENTION and TREATMENT

OF THE

## Disorders of Seamen and Soldiers in Bengal.

THE numerous obfervatiòns on the prefervation of the health, and the cure of
the difeafes of feamen, which have been publifhed by navigators of the medical profeffion as
well as others, have by no means exhaufted
this important fubjeét. They have undoubtedly fuperfeded the neceffity of long details,
or regular treatifes, which would unavoidably
prove, for the much greater part at leaft, a
mere repetition of former publications. The
information however to be procured from
books, refpeéting the health and difeafes of
Europeans in Bengal, is neither ample nor
fatisfaétory. One or two authors only have
written any thing material in this line; and
it were perhaps to be wifhed, as far as regards the treatment, that ftill lefs had ap-
B                          peared.

peared. In reality, the tendency, as well as the fcantinefs, of many of the obfervations which have been hitherto communicated to the public, affords not only the beft apology, but the beft argument alfo, for obtruding on the attention of the honourable company's medical fervants a few curfory remarks on this fubject, previous to the departure of the fhips for the enfuing feafon. It was hoped, that the prefs would have anticipated the neceffity of this addrefs. Although circumftances have occurred to retard the publication of a more ample communication; yet the reader of this paper may be affured, that no affertions fhall be admitted into it, except fuch as are founded on facts, and of this the prefs will foon furnifh undeniable proofs.

The fubject of prevention has a natural title to precedence. Under this head a few obfervations will be fuggefted on the fubjects of contagion, marfh effluvia, exceffes, drink, diet, cleanlinefs, exercife, ventilation, expofure to open air, pofture recumbent and erect, and inteftinal evacuations, with a concife recapitulation of the whole. Many of the remarks under the article of prevention will neceffarily prove applicable to the treatment alfo. Under the article of treatment, the diforders, which chiefly prevail in Bengal, and in

similar

fimilar latitudes, will be noticed in as concife a manner as the nature of the fubject will admit. The principal diforders in voyages to Bengal, or on fhore in tropical latitudes, are, fevers, dyfentery, affections of the liver, and fome other of lefs frequent occurrence.

## PREVENTION of DISORDERS.

## CONTAGION.

DURING the courfe of a pupil's medical ftudies at the univerfity, no part of his reading, or of the opinions of profeffors, makes fo forcible and ferious an impreffion on his mind as that fatal and impalpable agent in the excitement of fevers, which is called CONTAGION. On his firft entrance into practice he will expect to trace its footfteps wherever fevers or dyfentery appear, particularly in warm climates, which are fuppofed more favourable to its reception. A long voyage to India, and a long refidence in Bengal, quieted the writer's apprehenfions, and fhook his faith on this fubject. During the whole courfe of an affiduous practice there, he had not obferved, to the perfect conviction of his own judgment, either in or out of hofpitals, a

fingle

fingle inftance of contagion. Fevers and dyfentery were often epidemical, but never exhibited any appearance to his faculties, which could excite a fufpicion of contagion. Numbers of men afflicted with fevers, or with dyfentery, of the worft characters, were almoft every feafon fent from on board India-men to the hofpital at Calcutta; yet they were not thought to communicate contagion to the neighbourhood, to the attendants, or to the other patients. The Houghton Eaft In-diaman, on board of which the writer was favoured with the charge of the fick during the homeward-bound paffage, had fent a great many of her men, a very fhort time before it's departure, to the Calcutta hofpital. Thefe were deemed, by medical gentlemen who had recently arrived in India, to afford inftances of nearly the worft ftage of contagious fhip fever; yet it does not appear that this violent contagion had been communicated by thofe men to a fingle perfon in the hofpital, although it contained numbers ill of fevers and other epidemical diforders at the time. Where a confiderable number of perfons are feized with the fame complaint, in the fame place, and nearly at the fame period, contagion is gene-rally fufpected, particularly if the difeafe be febrile. During the voyage of the Hough-ton from Bengal to Madras, the fevers which prevailed

prevailed on board would certainly have been deemed fuch by moſt of the European faculty; and many of the circumſtances, according to the uſual modes of reaſoning and thinking, might have warranted the fuſpicion. He could not, however, perſuade himſelf that contagion had much agency in the general prevalence of the fever on board. It ſeemed natural, that the moſt perfect ſimilarity of diet, air, accommodation, conſtitutions of individuals, habits of life, &c. ſhould excite the ſame diſorder in numbers of different perſons independently of contagion. As the ſhip was deeply laden, the port holes and ſkuttles were obliged to be kept ſhut. The weather too was variable, often wet, and generally hot between decks; yet the number of patients who laboured under fever did not appear to increaſe to any conſiderable degree after we left the Broken Ground. This moſt probably would have been the caſe had contagion poſſeſſed much ſhare in the excitement of theſe fevers. While at anchor, the lower deck muſt have been better aired, and drier, as the port-holes could be kept open, than after our departure; and on this account the contagion, if any exiſted, ſhould have operated with more activity afterwards, in the rapid increaſe of theſe complaints.

B 3                    Change

Change of air indeed is on all occasions in India the moſt powerful enemy to obſtinate diſeaſes, particularly to fevers, but in the preſent inſtance the latitude alone, and not the air, was changed; certainly not the confined air between decks. *vide Errata Page.*

As every means in uſe to obviate or deſtroy contagion is equally applicable, and indiſpenſably neceſſary, in every inſtance of the prevalence of fevers or other diſorders amongſt a body of men, ſuch reflections on the ſubject might prove ſuperfluous, were not the general ſuſpicion of contagion prevailing in a ſhip of conſiderable inconvenience to the ſervice. On an emergency, when ſeamen might be wanted, they would naturally be deterred from engaging by ſuch fears; and an alarm of the kind, ſpread amongſt a crew, might tend to make them deſert the ſervice, diſpirit them in it, or render them at leaſt more liable, in reality, to be affected by the epidemic diſorders which happened to prevail at the time. With reſpect to thoſe already afflicted with fevers, an opportunity was afforded on board the Houghton to obſerve, that the terror attending the idea of a contagious diſeaſe has contributed much to that depreſſion of ſpirits, which often accompanies fevers, and renders them not only

more

more troublefome to the patient, as well as
the practitioner, but has often a very fatal
tendency in its own nature. The furgeon,
therefore, fhould be particularly cautious how
he encourages an opinion fraught with fuch
detriment to the fervice, as well as to the
crew, unlefs on the moft pofitive evidence,
which is poffibly not furnifhed by any combination of circumftances whatever that can
occur on board an Indiaman. Another inconvenience naturally refults from a fufpicion
of contagion in a fhip. The patients mefs-
mates, who are often found deftitute of humane
attention, to their comrades in ficknefs, will
affiduoufly avoid expofure to a fuppofed contagion, by attendance on their perfons, and
by rendering them innumerable little offices,
which, after all, may prove of more confequence in the cure of their diforders than
medicine itfelf.

A fhip under the ftigma of fuch a fufpicion could not be permitted to take in
refrefhments at the Cape of Good Hope,
fhould this be deemed eligible, on any occafion, as the fhip is vifited by a phyfician
from the fhore, to afcertain the exiftence of
any contagious diftemper on board. A furgeon, who was credulous with refpect to the
doctrine of contagion, would fcarcely think

B 4 himfelf

himfelf in confcience warranted to deny it, as long as a fingle cafe of the fever ftill remained on his lift. So prevalent were the apprehenfions of contagion on board the Houghton at firft, that the furgeon was frequently importuned, during the paffage from Bengal to Madras, by the paffengers and others, to declare whether the fever on board was contagious; and the prepoffeffion in this refpect was fo ftrong, that he was not by any means credited when he ventured to affure them of the veffel's exemption from infection. He was however fo well convinced of it himfelf, that he never thought of taking any precaution for the fecurity of his own health. The fick were always examined before breakfaft, with an empty ftomach and an open fkin, and with a conftant feverifhnefs, the remains of a former fever and affection of the liver. In this ftate of indifpofition, the fatigue attending the daily vifitation of the fick was completely exhaufting to the remains of ftrength. An hour's repofe, however, generally removed the great fenfe of fatigue and weaknefs; fo that, notwithftanding all thefe predifpofing circumftances, he had the moft unequivocal perfonal evidence of the abfence of contagion.

MIASMA.

# M I A S M A.

In the production of fevers, and fimilar difeafes on board of Indiamen, on their
firft arrival in Bengal, or during their term
of anchorage there, miafma, or marfh effluvia,
may be fuppofed to have a lefs doubtful
fhare than contagion. In the feafon of the
periodical rains, or at their termination, which
is perhaps the only unhealthy part of the
year in any place in India, the environs of
Diamond Harbour are very fwampy, and
much crowded with wood. Thefe circumftances alone might be deemed adequate to
the excitement of the difeafes in queftion;
but there are others, of even a more active
nature, which contribute largely to favour the
general ficklinefs of a crew. After a voyage
of confiderable length from Europe, during
which, although pofitive indifpofition may
not have been very common, a tendency to
fcurvy and other difeafes muft unavoidably
prevail. The elevation of fpirits, on firft
landing in Calcutta, and emancipation from
every reftraint of diet, &c. lead the people
into every poffible excefs in the gratification
of all their appetites. The very change of
air,

air, though poffibly for the worfe, has a powerful effect in raifing a temporary increafe of appetite for food, which is indulged to its utmoft extent. But this perhaps will be thought trifling in comparifon with the dreadful effects of ftrong liquors; from an early accefs to the largeft quantity of which it is impoffible to reftrain thofe who are permitted to land in Calcutta, or in its vicinity. Thefe fpirits are in general of the very worft and moft inflammatory kind. Few circumftances, perhaps, poffefs more activity in difpofing the people to diforders of every kind, than an unreftrained indulgence of the paffions with women. This generally encourages a total neglect of perfonal cleanlinefs, which, combined with profufe fweating from the great heat, and a fuppreffion of the infenfible perfpiration by an alternation or coincidence of hot and of wet weather, renders it a matter of great aftonifhment, not that fo many fhould be afflicted with fevers, but that any fhould efcape the operation of fuch powerful caufes.

## L I Q U O R.

It is with great diffidence that an individual can prefume to fuggeft a doubt refpecting

x                                    fpecting

fpecting the propriety of a practice which has received the fanction of years in this fervice. By fome unhappy influence of cuftom, failors view their drams as almoft the only object in exiftence worthy a failor's care. They eat their drams, and they drink their drams. Such is their attachment to this object of their affection, that they will remain for many days extremely ill, and perhaps dangeroufly fo, rather than be deprived of their liquor, by having their names enrolled on the fick lift. Nor is it very certain that any extenfive alteration in the regulation of their drams would be entirely exempt from troublefome confequences. It appears from the voyages of Captain Cook and others, that failors with the greateft reluctance admit of any alteration whatever in their eftablifhed habits and allowances. If any innovation prove a matter of fuch difficulty, where fubordination is fo complete, and where coercion has the fanction of law, it may be efteemed impracticable on board of Indiamen, where both are in a great meafure defective. The writer takes the liberty, notwithftanding, to declare his humble belief, that the daily allowance of liquor on board of Indiamen, in its moft favourable view, is at leaft unneceffary and ufelefs; that it gives the fhip's company

company a habit, which they might not otherwife have acquired; a habit which generally in the end deftroys more feamen, than are probably loft in any other way; that in cold climates, ftrong liquors are fo far from proving a protection againft cold and wet, the confequent inflammatory difeafes of the feafon, or even the fcurvy, that their ultimate effects are to furnifh a difpofition to be affected by all thefe. To fupply temporary ftrength, or fpirits for an occafional exertion, they may prove of fome ufe; but never without a fubfequent depreffion of both, in proportion to the previous excitement: that in lieu of promoting the cuticular difcharges, they have a manifeft tendency, perhaps both in their immediate and ultimate effects, to check them. In warm climates. they become ftill more deleterious. Their immediate action is to increafe the prevailing diftrefs from the heat; to induce or confirm a coftive habit, the fertile fource of moft of the difeafes of warm climates; to create a difpofition to be affected by bad air, by marfh effluvia, or by contagion, if fuch be fuppofed to exift. Thefe are their mifchievous effects, when taken in what is generally efteemed moderation, but when ufed to a degree of intoxication, which is very often the cafe with

feamen

feamen on fhore, particularly in Calcutta on
their firft landing, their fatal confequences
need not be enumerated. Reafoning alone
might afford a conviction of the juftnefs of
thefe fentiments; but thefe opinions are in
reality the refult of decided experience, and a
long attention to this fubject, during an extenfive practice amongft the foldiery of Bengal,
and occafional opportunities of obferving the
fame circumftances amongft the feamen of
all nations in the Eaft Indies.

Strong liquors, indeed, become lefs mifchievous in proportion to the quantity of water
with which they are mixed; and they may
ferve a good purpofe occafionally by mixture with water of a bad tafte, which fubftitutes the ftronger impreffion of fpirits on
the palate to the other. But whether, in
combination with water, they have ever any
confiderable tendency to correct the other
bad qualities of that element, is a fubject of
great doubt from every obfervation the writer
has had occafion to make in many parts of
India. Allowing them however their utmoft
ufe in mixture with bad water, yet it will
never be afferted that they improve either the
tafte or other qualities of good water. In
reality, they would prove of very unfrequent
ufe, were they allowed only in the inftances
of

of bad water, for water of this defcription
is not often found on board of Indiamen,
where provifions of every kind, efpecially this
very article, are always efteemed excellent.
In the vicinity of the various places where
Indiamen lay during their anchorage in the
mouths of the Ganges, the very beft water
may not be always procurable in any confiderable quantity, as the only provifion of
this effential article which the natives poffefs,
is the accumulation of the rain water of the
feafon in tanks conftructed for that purpofe;
but it has been always underftood, that the
moft abundant fupplies of water, taken from
the River Ganges beyond the reach of the
tide, are fent on board of all Indiamen.   No
water in the world can excel that of the
River Ganges, when collected and preferved
with due attention.   The water of Madras
and St. Helena, and, as far as my recollection
enables me to decide, the water likewife of
Bombay, are probably fuperior to any in Great
Britain.   One might venture to fuppofe that
fuch beveridge as thefe would fcarcely require
the aid of fpirits to render them either whole-
fome or palatable.

In the royal navy they feem very fenfible
of the mifchief refulting from the ufe of pure
fpirits; and it is accordingly their practice to
dilute

dilute the liquor before diftribution with a
very large proportion of water. This improvement, it is prefumed, might take place in the
honourable company's military, as well as
fhipping fervice, in which the daily allowance of fpirits is ferved out unmixed, and the
mixture left to the difcretion of a clafs of
people fo deficient in felf denial, that where
their own health and future comfort are concerned, they fcarcely appear endowed with the
inftinct of brutes. The additional allowance
of what is termed grog, on board of Indiamen, is perhaps feldom more diluted than
with equal parts of water, which ftill affords
a dram of confiderable ftrength. In reality,
were an allowance of fpirits in any inftance
indifpenfably neceffary to failors, foreign fhips
would labour under very great difadvantages.
This allowance is probably unknown to many
of them. An opportunity has occurred of
afcertaining the exemption from ficknefs
which foreign Indiamen enjoy. In one ftriking
inftance, during a voyage of nearly ten months,
one cafe only occurred of any diforder, a flight
intermittent fever. Towards the end of the
ninth month, after the crew had been for
fome time on a fhort allowance of provifions
and water, the fcurvy commenced its ravages;
but it was very extraordinary that it had not
made

made its appearance much fooner, as the
veffel had not touched any where for refresh-
ments, except at the beginning of the voyage
at Madeira. During the fhort paffage from
Madras to St. Helena, in the fineft poffible
weather, feveral of the Houghton's people
were afflicted with fcurvy; a ftrong and inftructive contraft!

## P R O V I S I O N S.

This immunity, which characterifes
the foreign commercial fervice, may be alfo
attributed in a great meafure to the more
fparing allowance of falted provifions on board
of their fhips, efpecially of animal food.
Farinaceous vegetables form the greateft part,
and indeed almoft the whole of the diet of
their people. This muft operate, not only
to the prevention of fcurvy, but of all febrile
diforders, particularly of that defcription
which is generally denominated putrid. The
allowance of falt meat on board of the honourable company's fhips is fo liberal, that
a failor may referve a very competent fhare
for every meal of every day in the week.
The other articles of diet appear very infipid
to palates habituated to the impreffion of

4                                              fpirits

fpirits and falt meat. They are confequently led to the ufe of their falt provifions at all their meals.

A vulgar error very generally prevails, that the allowance of falt provifions, as well as of fpirits, furnifhes ftrength for the performance of the fevere duty of feamen; but it does not require a detail of arguments to prove, that the article, which is beft fubdued by the powers of digeftion, and affords the largeft proportion of good nutriment, is that which fupplies the greateft ftrength and fpirits, and enables the people to undergo the greateft fatigue; nor can it poffibly be denied, that farinaceous aliment has greatly the advantage of falted provifions in thefe refpects.

It is eafy to conceive the benefit that might accrue from the fubftitution of other articles of diet to thofe of fpirits and falt meat, at leaft from a confiderable diminution of the latter in the warmer latitudes, where large quantities even of the beft and frefheft animal food are in the higheft poffible degree inimical to health, and difpofe, with an extraordinary power, to coftivenefs, to all bilious complaints, to affections of the liver, to dyfentery, and to the whole train of deadly diforders to which feamen are obnoxious in warm climates.

Providence feems to have directed the na-

C                              tive

tives of all warm climates in an especial manner to the use of vegetable diet; but its prevalence is by no means confined to them. It may be asserted with confidence, that the most athletic part of mankind, the peasantry all over Europe, excepting England, in some; measure live, thrive, and grow strong on a diet almost entirely exclusive of animal food, and that probably two thirds or more of the inhabitants of the globe live for the most part, or entirely, on vegetable food. Those uncivilized nations are thinly populated where vegetation is not abundant, and where animal food affords the only source of half-starved existence.

## CLEANLINESS.

SUCH are the most powerful causes of the diseases of seamen. But another exists scarcely of less activity, and of a prevalence as general as the former. If inattention to cleanliness prove a source of scurvy and other complaints in cold climates, what must its pernicious tendency be in warm regions, where the cuticular discharges are so much more profuse and acrid; of course their retention must become much more detrimental.
The

The falutary effects of unremitted attention to the perfonal cleanlinefs of a crew have been evinced by the experience and teftimony of many navigators. The powers entrufted to commanders in the royal navy enable them to inforce every regulation, which their prudence may deem neceffary for the prefervation of a healthy crew; but the cafe is far otherwife on board of Indiamen, where a commander might expofe himfelf to the vexation of eternal profecutions on his arrival in England, were he to exercife that authority, which is abfolutely neceffary to compel the people to pay any attention to this indifpenfable prefervative of health. Perhaps there is not a clafs of perfons in exiftence fo devoid of every attention to cleanlinefs as Britifh failors, when abandoned to their own guidance. The foppery of fome foreign feamen has a moft ufeful tendency in this refpect; but a true Britifh tar would feldom, I believe, dream of a change of linen, were he not occafionally compelled to it by being wetted to the fkin; their heads remain long ftrangers to the comb; their beards to the razor, and their faces to cold water. The only bath they ufe is that which heaven fends them occafionally in heavy fhowers.

It is not to be doubted that the fea-water

bath,

bath, independently of its other excellent
effects, which are allowed to be numerous,
would prove highly ufeful with refpect to
cleanlinefs alone. The general ufe of the
cold bath amongft the natives of warm
climates, more general even than the prevalence of vegetable diet, both of which have
become objects of religious obfervance in
many places, affords unequivocal proofs of
its great utility in the prefervation of health.
Sea-water has not probably a decided fuperio-
rity for the purpofes of bathing over any
other kind of cold water.

If inftinct alone can teach the brute creation
thofe means of felf prefervation, which are
beft adapted to the regions they inhabit,
it cannot be fuppofed that the general prevalence of vegetable diet and frequent bathing
amongft rational creatures in tropical countries is the mere effect of accident, and not
the refult of reafon and experience, operating
flowly perhaps and almoft imperceptibly, but
with ultimate certainty, on the habits of nations, in the earlieft conditions of fociety;
but neither the prevalence nor utility of
bathing and vegetable diet is by any means
confined to the natives of warm climates;
to them indeed the habitude of fuch practices would feem more effentially neceffary;
but

but they cannot be deemed ufelefs even in the frozen regions of the north. The greateft and moft fudden changes of weather can fcarcely be expected to affect, even in a trifling degree, the perfons of Ruffians, who can plunge from the warm bath into water whofe temperature is nearly at the freezing point.

Circumftances of fuch importance may poffibly merit the attention of the honourable court of directors. Commanders, if that degree of regard to the fubject be deemed expedient, might be furnifhed with orders and powers to compel the feamen to wafh their whole perfons once or oftener during the courfe of every week. This has been found of difficult execution, even when pre-fcribed medicinally by the furgeon; and there is not perhaps any medicine, the moft naufeous, but what they would prefer to it. They would almoft rather take a pint of falt water internally than externally.

# EXERCISE.

INDOLENCE would fcarcely be expect-ed to prevail in a fhip, where it is generally prefumed the duties of the veffel afford fea-

C 3                                     men

men conftant or fufficient occupation. This
in reality may be the cafe in voyages to colder
climates, on more boifterous and inconftant
feas, and in more variable weather, during
which inceffant work and fatigue may prove
as inimical to the health of a crew, as
the moft torpid inactivity ; but in voyages
to the Eaft Indies, during the prevalence
of trade winds, and nearly on all occafions, a
fhip's crew may indulge in very enervating
habits of lazinefs, and it requires the encou-
ragement, perfuafion, or compulfion of authority, to induce them to take the degree of
exercife which in all fituations would be
deemed effential to the prefervation of health.
In this refpect we might fubmit to be taught
a leffon by the French, and other foreign
nations, who practife thofe means of exercife
on board of their fhips, that contribute as
much to hilarity of mind as to perfonal vigour. Dancing is the moft obvious and the
moft general of thefe, and this, or any other
modification of exercife, may be deemed fo
much the more neceffary for Britifh failors,
as their diet is infinitely groffer than that of
foreigners, and renders them more obnoxious
to the depreffion of animal fpirits, which
favours the approach of difeafes, as well as to
the difeafes themfelves, from the immediate

5

effects

effects of fuch aliment. The various modes
of exercifing the people is neceffarily referred
to the difcretion and ingenuity of the commander and his officers. It is very well
known, that this cuftom has been introduced
by fome commanders in the honourable
company's fervice with very obvious benefit
to their crews, and there is ample teftimony
in its favour in the printed voyages of various
navigators.

The exercife of dancing is one great in-
ftrument of the prefervation of the flaves on
foreign fhips. Their commanders of veffels
employed in the flave trade are moft puncti-
lioufly attentive to it. Ocular proofs have
occurred to what a degree their endeavours
are crowned with fuccefs. Their attention
alfo to the perfonal cleanlinefs of the flaves,
as well as to the neatnefs of their accommoda-
tions, is truly admirable, and probably much
fuperior to what is practifed on board any
mercantile fhips whatever. It is not exaggeration to declare, that every part of the deck
appropriated to the accommodation of the
flaves is kept as clean as an officer's cabin
on board of an Indiaman. It will be found
that exercife and cleanlinefs are generally
affociates; that the one naturally leads to the
other, and is feldom neglected, unlefs where

the

the former is deficient. Exercife may not only be confidered as a grand prefervative of health; but in gentler forms, as a very material agent in its reftoration. I fhall however refer the confideration of it in this point of view to the fubfequent article.

## VENTILATION.

THOSE leading circumftances in the prefervation of the health of feamen, which appear moft obvioufly ufeful, and moft within the fcope of practice, have been noticed in the preceding pages. We may now proceed to fuch as are by no means lefs effential, but not always equally practicable. Of thefe, ventilation, including expofure to the open air, is of the firft importance.

A conftant ventilation between decks would prove by far the beft fecurity againft difeafe. But this is not very eafily procured to a proper degree. Indiamen are fometimes fo deeply laden that the port-holes cannot be opened even in moderate weather, and never perhaps when the fea is rough. Although the air be not admitted with fufficient freedom, yet the water oozes in great abundance. The deck has therefore leaft ventilation

ventilation when it requires its affiftance
moft, to evaporate the dampnefs occafioned
by the water. On all occafions when it may
be deemed neceffary to fhut the port-holes,
it ought furely to be executed in fuch an
effectual manner as to exclude the water
completely, if poffible. This, it is imagined,
is feldom the cafe. But independently of the
dampnefs from the admiffion of fea-water,
the people fhew much negligence in wet-
ting their *births*, or the fhare of the deck
allotted to the ufe of each mefs. The attention of officers might undoubtedly obviate
this inconvenience, by enquiring on all occafions into the caufe of the naftinefs, and
inflicting proper punifhment, fuch as a ftop-
page of drams, on the authors. Each mefs
might eafily be compelled to keep their al-
lotment of the deck very dry, and clean in
other refpects. Water fhould never be allowed to remain on the lower deck for a moment, although the fwabs fhould furnifh
employment for a great proportion of the
fhip's crew. In the worft weather, a perfon
might fuppofe the deck would be kept pretty clean by thefe means. It fcarcely comes
within the writer's province to obferve, that
it would be fortunate if the port-holes of
the honourable company's fhips were fo
much

much above the level of the water as to admit of being opened on nearly all occasions during a voyage to India. A deck crowded with chests and hammocks must neceffarily prove very unfavourable to ventilation. The latter fhould never be allowed to remain below in fair weather; and as few as poffible of the former fhould be admitted. In reality, the people themfelves, unlefs at the periods of regular reft and meals, fhould be compelled to remain on deck as much as poffible, except during rain; nor fhould this be omitted, even in very hot weather, when they can enjoy the protection of an awning. Indifpofition itfelf fhould not exempt them from conftant expofure to the open air. A feaman, when his name is once enrolled on the fick lift, thinks himfelf entitled, if inclined, to lay in his hammock, or on his cheft, or at leaft to remain below, all day. Nothing can be more pernicious than fuch an unwarrantable indulgence to the fick. The error in this refpect is very prevalent, and perhaps fatal, even in the royal navy. A part of the deck is allotted as an hofpital for the fick, furrounded in general by canvafs. Although every attention is beftowed on its cleanlinefs, it is doubtful whether the greateft feverity of weather would prove fo

detrimental,

detrimental, even in the worſt caſes, as allowing the ſick to ſwing perpetually in their
hammocks in ſo confined a ſituation. Under alarming (may we venture to call them
for the moſt. part imaginary) apprehenſions
of infection, the collection of the ſick in one
ſpot, and their ſeparation from the reſt of
the crew, may be deemed a meaſure highly
prudent, independent of its expediency with
reſpect to attendance and care of every kind.
But what object of convenience, or of a
ſalutary tendency can an encloſure of canvaſs
effectuate, even in the worſt circumſtances
of diſeaſe or infection? Can there be a variety of opinions on the deleterious effects
of a momentary confinement, on ſuch air?
eſpecially when the more feeble patients are
indulged with the uſe of the bucket, which
cannot be removed with ſufficient expedition
to prevent its contaminating in a conſiderable degree the unwholeſome atmoſphere
within the canvaſs. Whatever the diſorders
may be, which happen to be impriſoned in
this manner, they cannot fail to be much
aggravated. But if unfortunately dyſentery
prove of the number, what muſt be the
reſult of a frequent recurrence to the bucket,
and of the duration of efforts, productive in
reality of ſmall but highly vitiated diſcharges.

charges. Under such circumstances, will the renewal of air be sufficiently rapid to afford protection to the sick against the operation of the putrid effluvia.

If such a formidable host of objections arise to the confinement and enclosure of the sick in the predicament already described, what motives of utility can be suggested in favour of the plan, when the terror of contagion is not suspended over the imagination, and when the greatest part of the sick will be found capable, on a spirited trial, of attendance on the surgeon, instead of requiring his very imperfect attentions in their hammocks? A person might naturally suppose, that nothing could have a more salutary tendency in the revival of a patient's spirits, than a participation in the cheerfulness, and as much as possible in the amusements, of their companions in health, and that few circumstances, on the contrary, would contribute more effectually to depress the spirits, and to aggravate the distemper, than the exclusive society of the diseased, the dying, and the dead.

We may suspect that cases do not exist, at least the writer has not witnessed any, which require the indulgence of the hammock, wherever it may be permitted to hang. In the

the laſt ſtages of expiring exiſtence, whatever its inutility may be on other occaſions,
it would be cruel, as it is impoſſible, to refuſe
it; but in circumſtances of leſs poſitive neceſſity, even in thoſe febrile diſorders which
are characteriſed by the greateſt ſenſe of debility, the moſt particular care ſhould be taken to compel the ſick to expoſe themſelves
to the open air, and to uſe as much motion
as their condition will allow.

This degree of gentle motion may naturally enough come under conſideration in the
preſent article. In theſe diſeaſes it will be
found, that the power of motion increaſes
with the frequency of the attempt. During
the voyage of the Houghton from Madras to
England, many, and perhaps all of the ſick,
afflicted even with fevers, in the intervals of
the paroxyſms, were perſuaded to do more or
leſs duty on deck, and there was reaſon to
ſuppoſe, with conſiderable benefit to their
complaints. This attention to ſome degree
of exerciſe, as long as a muſcle in the body
retains the power of motion, is univerſally
allowed to be of the utmoſt conſequence in
retarding the worſt ſymptoms of ſcurvy. It
does not appear that a ſufficient reaſon exiſts
for its prohibition in many other diſorders.
That ſet of ſymptoms, which are generally
denominated

denominated the *nervous fever*, is acknowledged to exhibit the moſt frequent as well as
the moſt fatal form of fever which prevails
in ſhips, yet the enlightened experience of a
phyſician, who publiſhed a treatiſe on the
nervous fever, has placed it beyond the poſ-
ſibility of doubt, that gentle motion and broad
expoſure to the open air prove the moſt eſ-
ſential requiſites in the treatment of this diſeaſe.

Directions, which recommend exerciſe
or motion, ' as far as the patient be capable
of uſing them,' are too vague to be of much
uſe; for a ſailor, who is diſpoſed to indulge
habits of indolence, will ſcarcely ſcruple to declare that he has not ſtrength for the purpoſe.
In reality they are themſelves greatly deceived
in this particular. In theſe, and nearly in all
diſorders, the patient is frequently oppreſſed
with ſuch a ſenſe of debility, that he would
think the practitioner very unreaſonable, if
not mad, who ſhould attempt to perſuade him,
that his feelings deceived him; that the
weakneſs was not real or permanent; and that
a reſolute perſeverance in gentle exerciſe, and
expoſure to the open air, would tend greatly
to remove it. A perſon in the higheſt vigour
of health will experience ſome inconveniences,
a degree of languor, a little nauſea, and perhaps

haps a flight head-ach, at leaft in warm climates, if he prolong in the morning his imperfect flumbers in bed for two or three
hours beyond the ufual time; but if the
fame perfon fhould extend this indulgence to
the expiration of twenty-four hours, I may
venture to affert, that two days would fcarcely prove fufficient to reftore him to the fame
fenfe of all his faculties which he poffeffed
before.

If a mind and perfon the moft healthy
fuffer fo readily by too long indulgence in an
horizontal pofture, is it confiftent with common fenfe or with reafoning to fuppofe, that
a man already under the preffure of thofe
fymptoms, perhaps in their worft ftages, fhould
not only not fail to be injurioufly affected
by it; but fhould even derive much benefit
from a circumftance which would deprive
him of ftrength, if he poffeffed it in a
healthy degree. There is certainly an abfurdity in the fuppofition; but it refts not altogether on fuppofition; for the writer may
truly declare that the whole courfe of his experience in Bengal, as well as on board the
Houghton, militates ftrongly againft all confinement to a bed, an horizontal pofture, and
even to a room or a lower deck; moft per-
aps in the diforders in which the ability to
abandon

abandon thofe indulgences is leaft apparent
to the practitioner or the patient.

During the progrefs of the Houghton's
voyage, it was a conftant endeavour to deprive
the fick of every defcription of the ufe of
their hammocks in the day-time, unlefs while
the violence of a febrile paroxyfm lafted, and
even on thefe occafions exhortations were not
fpared to perfuade them to quit their hammocks, and remain as much as poffible in an
upright pofture, from a thorough conviction
that although an horizontal pofition might
prove lefs uneafy to them at the time, yet
the violence of the fit would terminate fooner, and leave much lefs affection of the head
and ftomach, as well as lefs proftration of
ftrength after it. Neither exhortations however, nor threats, ufed to avail in many inftances, without recourfe to the authority of
the officer of the watch to compel them to
relinquifh their hammocks, with the rifk of
being rendered obnoxious amongft the people
to the imputation of inhumanity. The fymptoms which feem to refult from indulgence
in a recumbent pofture, efpecially in a confined atmofphere, are principally a languor or
greater fenfe of debility ; a perpetual drowfinefs, but no refrefhing fleeps ; a great increafe
of heat, or excitement of profufe and pernicious

cious perſpiration; a decay of appetite; a
coſtivenefs, or retention of ſtools and urine;
a fenfe of oppreſſion about the præcordia, or
accumulations in the cavity or the vicinity of the ſtomach, liver, and other bowels;
a naufea; a weight, pain, and particularly giddinefs of the head on riſing, accompanied
with a fuffuſion or muddinefs of the eyes,
and even of the ſkin, which are pretty clear
proofs of a morbid abforption from torpid
bowels. Thefe are pretty certain confequences of confinement for any length of time to
an horizontal poſture, even where the previous indifpoſition has been very trifling; but
we may repeat, that they are alfo the very
fymptoms which characterife moſt fevers,
and particularly the worſt. If the indulgence
in queſtion be deemed ufeful during the prevalence of thofe fymptoms, it will be neceſſary to allow that a caufe not only does
not produce its natural effects, but that it
operates to the removal of thofe effects when
excited by any other caufe. To the writer's
comprehenſion this appears to involve a mon-
ſtrous abfurdity. It may be prefumed that
very little doubt will remain, after what has
been ſtated, of the impropriety attending this
miſtaken indulgence to the fick, merely as
it regards their own recovery, independent of

D                                        its

its pernicious effect on the air below, and consequently on the rest of the crew, or at least on their more healthy messmates. To them it must always prove a great inconvenience, if not a positive mischief. It is in reality a general nuisance, but in a most particular manner to the surgeon and his assistants. A surgeon may often acquire more knowledge of the state of his patient from a view of his countenance, and his general appearance, than from the most distinct answers to the most sagacious questions he may propose. Much information is to be procured from seeing the patient sit, stand, or walk.

This intuitive knowledge, which a practitioner receives from the general appearance of a patient at first view, affords him perhaps more light in the treatment, than the utmost exertion of reflection and reasoning on the more obvious symptoms. It is here where a practitioner's sagacity and experience appear to operate imperceptibly even to himself. A surgeon, who is obliged to visit a number of patients in their hammocks, cannot enjoy the benefit of such instructive impressions, nor can his attention to each case be so minute in other respects as he could wish. Daylight, as it frequently has no access, can furnish no assistance; and the appearance of a patient in

his

his hammock by the light of a lantern is extremely different from what it would be on deck in broad daylight; nor can the furgeon, under fuch circumftances of inconvenience, particularize at the greateft length on his diary every trifling appearance which can tend to throw light on the cafe. Every patient, therefore, who can poffibly move his limbs, fhould be compelled to quit his hammock and his birth at the hours of attendance, and to wait on the furgeon in the part of the fhip appointed for that purpofe. It may be affirmed once more, that notwithftanding every appearance of debility, a patient will fcarcely ever be entitled to the enjoyment of his hammock in the day-time.

## INTESTINAL EVACUATIONS.

MANY of the preceding obfervations are of more importance in the treatment than in the prevention of diforders; but they will be found ufeful in both. The fcene, of which the writer became an anxious obferver on board the Houghton, has convinced him that it is of the utmoft poffible confequence to confider every means which may tend to obviate the dreadful effects of an unhealthy

D 2                    neigh-

neighbourhood, in an unhealthy feafon, on the feamen of the honourable company's fhips, on their arrival at Diamond Harbour in Bengal. It is fcarcely neceffary to notice, that the arrivals often occur during the periodical rains, or immediately after their ceffation. It is much to be regretted that the tafk is eafier, to fuggeft than to enforce a compliance with the neceffary precautions. That there do in reality exift fuch as will feldom fail to prevent the fatality of difeafes refulting from thofe fituations, we may be perfuaded. Many of them have been already enumerated; but one of the very firft importance ftill remains, which to many will appear infignificant or ridiculous, although it undoubtedly affords by far the moft certain protection againft all difeafes in warm climates, efpecially fuch as perfons on their firft expofure to an unhealthy fituation are moft likely to experience. Early inteftinal evacuations on the flighteft approach or fufpicion of indifpofition, or previous even to an intimation of this kind, will, I may almoft fay infallibly, procure an immunity from danger, if not from difeafe. However obnoxious to ridicule a propofal of this nature may feem, we need not hefitate to declare, that a general application of thefe means to a whole

x                                      crew,

crew, immediately before their arrival in harbour, and a frequent repetition afterwards,
would be attended with the moſt ſalutary
conſequences. It is not the writer's province
to judge how far this may be practicable in
the honourable company's ſhips. Sailors
would univerſally think the propoſer of ſuch
a precaution inſane. But as long as men indulge in exceſſes of diet, or do not reduce
confiderably the quantity of their uſual food
in thoſe ſituations, the frequent neceſſity of
inteſtinal evacuations is not to be doubted.
Amongſt the officers, however, there are rea-
ſonable men, to whom this precaution is recommended in the ſtrongeſt terms, with a
poſitive aſſurance, that they will never be
deceived in the confidence they may repoſe
in it; for they muſt be of an original con-
formation, widely different from their fellow
creatures of the military eſtabliſhment in Bengal, if they be not affected by precautions,
which, during the whole courſe of an attentive experience, have uniformly ſucceeded
with the latter.

The unhealthineſs of Diamond Harbour,
and perhaps of every part of the mouths of
the Ganges in the periodical rains, or more
properly about their termination and the
commencement of the cold weather, is to be

D 3                                    greatly

greatly lamented, but, I fear, not remedied. However, more ſtreſs may poſſibly have been laid on the effects of an unwholeſome vicinity, conſidered abſtractedly, than they merit, if it be allowed that there do exiſt real means of obviating to a certain extent thoſe effects, and that conſequently the miſchief is generally as much the fault of inattention as of the climate.

## RECAPITULATION.

WE may now review, in as conciſe a manner as poſſible, the principal precautions which may be uſed to obviate the baneful effects of an anchorage in Diamond Harbour, at the ſeaſon of the uſual arrival of the honourable company's ſhips. The following are the moſt material. A very conſiderable reduction of the quantity, and alteration of the quality of their food, which ſhould conſiſt as much as poſſible of vegetables, ought immediately to take place. Theſe are eaſily procurable in Calcutta and the environs, to any extent of demand, in the ſeaſon in queſtion. European vegetables may not always be found in ſufficient quantity; but the eſculent plants of the country are, in the eſtimation

eſtimation of many, by no means leſs ſalutary, or even leſs agreeable to the palate. The honourable company's hoſpital at Diamond Harbour, which was erected at a large expence, has not as yet proved of much utility to their ſervice; but a garden for the proviſion of freſh vegetables, on a very adequate plan, in the ſame neighbourhood, would demand no very conſiderable ſum; and might prove of infinite advantage. It is proper to obſerve that vegetables, which are peculiar to the ſoil, require very little culture in thoſe ſeaſons.

With reſpect to drink, we ſhould not doubt the univerſal ſuperiority of pure water over all other beverage in every ſituation. This article, with ſome trouble indeed, may be always procured of a good quality, and eaſily cleared from the muddineſs by ſprinkling the ſmalleſt quantity of powdered allum into the jar or caſk, and allowing it to remain at reſt for a day or two. It would ſhock the prejudices of people greatly, to aſſert that wine or ſpirits, taken in great moderation, muſt neceſſarily prove hurtful; but it may be ſafely aſſerted, that the leſs of any fermented liquor which is uſed, the greater probability there will be of preſerving health in Bengal. One plauſible argument occaſionally ſug-

geſted

gested in favour of the continuance of the allowance of drams, not only to the sailors in
general, but likewise to those whose names
are enrolled on the sick list. It seems that
drams are the current coin of the ship, particularly at sea. Sailors, destitute of reflexion,
and slaves to their senses, are scarcely sensible
of the value of any object which has not an
immediate use. A dram is consequently of
more consideration to them than a piece of
coin, which would procure them a number
of drams at a remoter period; nor is coin
of any kind, at least of the smaller denominations, very common on board of a ship.
Hence every little office is repaid in this currency; and as the sick stand most in need of
the assistance of others, they should not be
deprived of the means of purchasing or re-
warding it. The tendency, however, of this
practice would naturally be, that the most
useful persons on board would soon become
the greatest drunkards, and in a short time
consequently the most useless, as their sup-
plies of liquor would be unlimited. This argument therefore operates decidedly against
the allowance of drams, at least to the
sick.

The next article recommended for the
preservation of health, is the most particular
attention

attention to cleanlinefs in all its branches.
This obvioufly includes a conftant ventilation
and frequent fcouring of the decks, as well
as a prohibition againft the number of chefts,
which are generally allowed to crowd the
lower deck. The hammocks fhould never
be permitted to remain below during the day,
if it do not rain, and the people fhould be
occafionally compelled to wafh them. Soap
might be furnifhed to them for this purpofe;
for what is generally called Bombay foap
lathers well with falt water, and is extremely
reafonable in price. Clean clothes are perfectly
in the power of the pooreft individuals in Bengal, as well from the cheapnefs of the materials as of the wafhing; but the moft effential
article of cleanlinefs is fortunately alfo the
moft acceffible. The general bath is attended with advantages, even independent of
its cleanfing effects. Seamen fhould be compelled to wafh their whole perfons as often
as poffible, and officers will experience the
advantage of its daily ufe. Authors who
have written large treatifes on the difeafes of
warm climates, from the warm climate of a
chimney corner in Europe, have given the
moft pofitive prohibition againft the ufe of
the cold bath in the prickly heat and other
cutaneous eruptions; but thefe are the mere
effufions

effufions of fpeculative men, and deferve no manner of credit. The cold bath fhould not be interrupted on this account. It has not been obferved of the fmalleft difiervice in great numbers of inftances, under every variety of conftitution, age, and fex.

No further notice may be taken of the other heads of prevention, except that of inteftinal evacuations, which is by far the moft important of any. Thefe indeed, even as prophylactic, muft often be ufed to an extent that would furprife European practitioners, if given in thofe diforders in which they are by them deemed moft neceffary. Caftor oil, or a combination of calomel with refin of jalop, fcammony, cathartic extract, and other purgative medicines of a fmall bulk, are beft adapted to this purpofe.

In the preceding pages, the writer has endeavoured to fuggeft fuch hints refpecting the prefervation of health in the honourable company's fhips, as a long experience in India, joined to the opportunities which occurred during the Houghton's voyage, enabled him to communicate. He is fenfible that fo fmall a fhare of experience in the fea line would give but little claim to attention, if a refidence of nearly ten years in India, and a conftant employment in profeffional duties during

during that period, may not be thought to counterbalance in some meafure the difadvantage. Under the article of prevention, the obfervations of others may poffibly have been fometimes repeated ; but the remarks that may be made on the treatment of the moft frequent diforders in Bengal, will contain few opinions perhaps, but feveral modes of practice, of which the faculty are lefs generally apprifed. Some of them muft neceffarily appear of a nature the moft extraordinary to European judges, as well as to moft medical gentlemen in India ; and if the decifion on their merits be abandoned to the former, they will infallibly experience an immediate condemnation. He deprecates a trial by incompetent judges, and fubmits the merits of the caufe to the determination of the tribunal of facts alone.

To oppofe, or even to controvert, the doctrines and the practice of the moft eminent writers, teachers, and practitioners of the prefent age, is a fervice of danger, in which no man whofe livelihood depends on the extent of his practice, and confequently on the recommendation and protection of his fenior brethren, will venture to engage. The attempt would prove vain, and ruinous perhaps to his future fuccefs in Europe. The flighteft eſt puniſhment would probably be the deriſion of the faculty, and the neglect of all the world. But the happy predicament of the faculty in Bengal, which renders their income independent of the opinions of their brethren, or of the public at large, allows them a liberty of thinking and talking on profeſſional ſubjects, and gives ſcope to the exertions of their own judgment in deviations from common practice. Theſe are advantages which the former do not poſſeſs, or dare not exerciſe, under the dreadful apprehenſions that any ſeceſſion from common routine might expoſe them to the imputation of raſhneſs, and operate to their utter excluſion from lucrative practice.

With ſuch advantages in favour of medical gentlemen in that country, it is very ſurprizing that more improvements have not originated amongſt them. It is true, they firſt introduced the proper treatment of obſtructed liver; and although this improvement is not of a very recent date, yet few practitioners in Europe, and thoſe perhaps only in England, have as yet acquired any diſtinct idea of the nature of that diſorder, with the preſent mode of treatment.

The proper management of fevers, and of other diſeaſes, is not perhaps equally general; for

for there are, no doubt, fome medical gentlemen, both in the honourable company's fettlements and fhips, whom judgment and experience have not as yet taught to relinquifh
the doctrines of univerfities, and the practice
from books fo little calculated for any meridian but their own, in favour of means that
are obvious, fimple, and natural.

# F E V E R S.

A VERY few general obfervations fhall
be firft offered on the fevers which generally
occur in the honourable company's fhips, and
in Bengal; next, fome curfory remarks on
the principal medicines in ufe amongft their
medical fervants; and conclude with a concife
ftatement of the method of cure, which has
feldom, if ever, on a fair trial, failed.

Doctor Pafly, at Madras, was probably the
firft who ventured to confide in his own obfervation, and to deviate from the deftructive
practice of the times *. The few who have
had opportunities of obferving the methods

* The name of another gentleman of eminence was
originally introduced here, but is now omitted, as the
author was informed he had expreffed diffatisfaction on
the occafion.

which

which this gentleman purfued with fuccefs,
are not fufficiently numerous, active, or com-
municative, to afford them general currency.
Two years ago, an improper method of treating fevers was thought to prevail in many
parts of the country, not altogether from a
neglect of the proper means, but in general
from a timidity, which deterred gentlemen
from the exercife of thofe means to an extent
that would have enfured fuccefs. Relin-
quifhing the tafk of fubverting old, and efta-
blifhing new theories, it may be afferted in
general, that the ideas entertained of the origin of fevers in warm climates, at leaft in
Bengal, are probably erroneous; that the
truth of this affertion will principally appear
from a mode of treatment which has been
attended with invariable fuccefs; that, to the
degradation of fpafm, and other ingenious
hypothefes, with the practice founded on
them, the order of pyrexiæ, under the general
denomination of fever, may be deemed uni-
verfally to originate, in thofe latitudes, from
the bowels and their contents; that the indications fhould confequently arife from this fource,
and the curative means be derived from fuch
medicines as operate on thofe parts by evacuation or otherwife, but particularly by *purging*.

As

As the fymptoms, amongft nofological writers, conftitute the difeafe, and thefe as effects are not generally removable without the previous removal of their caufe,— we muft fearch for the caufe of thefe fymptoms in the circumftances which attend their removal. If the circumftances which accompany or effect the removal of diforders, according to prevalent notions, diametrically oppofite, be exactly fimilar, it is reafonable to conclude, that the caufe which thefe circumftances evince muft be the fame, notwithftanding the variety and apparent contraft of the effects. If the removal of all that variety of fymptoms, which have been fuppofed to conftitute the diftinct fevers, inflammatory, putrid, nervous, &c. has been attended by the fame circumftances, and effected by the fame means, it will be equally reafonable to infer, that the caufe of all thefe fevers is one. That inftances of fuch appearances as are faid to attend different fpecies of fevers have originated from fimilar caufes, have yielded to fimilar means, and that thefe means have been chiefly purgatives, a comparifon of a large collection of cafes, which have occurred in Bengal, and on board a fhip, with thofe appearances, will prove to the fatisfaction of every perfon, whofe judgment is unbiaffed by prejudices

prejudices acquired at the univerſity, or in
the ſhop, or by the reſpect which is due in a
certain degree to great names.

It ſhould not prove a ſubject of wonder,
that bowels oppreſſed by vitiated bile, mucus, or other offenſive matters, muſt neceſſarily occaſion ſuch a variety and direct contra-
riety of ſymptoms, when it is acknowledged,
that an affection of thoſe parts is capable of
exciting the motley tribe of ſymptoms which
exiſt and ſucceed one another in the hypo-
chondriac diſeaſe.   There cannot be a doubt,
but that the ſame diſorder may diſguiſe itſelf
under appearances exactly the reverſe of one
another.   The whole train of what are uſu-
ally called nervous affections ſtands in direct
proof of this, and affords the ſtrongeſt ſuſpi-
cions of a derivation from a ſource ſomewhat
ſimilar to that of fevers.   The venereal diſeaſe and the gout perſonate a variety of diſorders; the latter would alſo ſeem to have
ſome connection with the cauſe of febrile
complaints.   Many arthritics of various de-
ſcriptions have fallen under my obſervation,
but in no one inſtance without thoſe ſymptoms which characteriſe vitiated accumulations in the bowels.   In whatever manner
the gout of Bengal be ſuppoſed connected
with appearances of prevailing bile, whether

as

as caufe or effect, it has never occurred to the writer, on any occafion, exempt from an obvious connection with that fecretion; and it will be found that, in thofe climates at leaft, the proper management of gout fully warrants this judgment refpecting its nature. Even in England opportunities have occurred of obferving fymptoms, which have been deemed gout by attending phyficians, vanifh immediately after very large evacuations of bilious fordes had fucceeded the operation of purgatives. All the complaints of children, without an exception, have either the moft intimate connection with, or derive their very exiftence from, the bowels; and poffibly a fimilar connection, with a fimilar fource of exiftence, pervades moft of the diforders to which every age and fex are obnoxious.

When cafes, therefore, occur in thofe regions, which exhibit the appearances of an inflammatory, or of a low nervous fever, fhould the phyfician, according to the beft knowledge he may have acquired from lectures or from books, pronounce them diftinct diforders, and oppofite in their nature and treatment, the patient would, in general, have a very unfair chance for his life. But the error is more fatal in the latter, in which fudorifics, ftrengthening medicines, and cordials,

E                                    dials,

dials, are generally prefcribed. In the former, indeed, inteftinal evacuations are allowed,
in a limited manner, by practitioners in Europe.

An opinion generally prevails, that the difeafes of warmer latitudes differ very materially
from fuch as afflict the inhabitants of cold
climates, and that the methods of treating
them fhould confequently vary; under this
impreffion, the beft practitioners in India
have ventured to deviate in fome meafure from
the practice of Europe, or have rather exer-
cifed the means fometimes recommended by
authors to a greater degree.

Few medical gentlemen, unlefs on their
immediate arrival in the Eaft Indies, confine
inteftinal evacuations, at the commencement
of many diforders, particularly of fevers,
within the limits of European practice; but
fewer ftill poffefs experience and courage to
exert thofe means with the energy which is
abfolutely neceffary for the prefervation of a
patient on many occafions.

Authors have recommended more confiderable evacuations in fevers purely bilious, than
in thofe of a putrid, nervous, or inflammatory
character. A gentle vomit, and a laxative,
perhaps one repetition of thefe with occafional glyfters, conftitute the whole of the

x                                evacuations

evacuations from the ftomach and inteftines;
but in cafes fuppofed to be of the true bilious
kind, thefe evacuations, though procured by
the gentleft means, are recommended to be
repeated oftener, and prolonged, perhaps, until an intermiffion or a remiffion take place,
when the bark is exhibited without lofs of
time, or a fcruple refpecting the quantity, to
obviate a return of the fymptoms, but in
reality a recovery from the difeafe. When
the nature of the diforder is very obvious in
bilious fevers, moft individuals of the faculty
will not hefitate to promote thofe evacuations
to a degree beyond European practice; and
the means are only defective in celerity and
vigour; ftill, however, with a prejudice in
favour of the bark in the firft, or amongft
the moft intrepid and intelligent in the fubfequent remiffions or intermiffions of the
fever. But when the bilious fever is dif-
guifed under doubtful appearances, or, to
fpeak more properly, when the foul contents
of the ftomach and inteftines excite appearances which perfonate the inflammatory, the
putrid, or the nervous fever, and their feveral
modifications, the evacuations are generally
reftricted to a vomit and a laxative medicine,
perhaps a fingle repetition of the latter with
occafional glyfters, fucceeded by diaphoretics,

E 2				vigorous

vigorous antifeptics, corroborants, cordials, ftimulants, opiates, and death! A reference may be made to the diaries of hofpitals, and to the journals of furgeons of the honourable company's fhips, for incontrovertible proofs of the reality and frequency of thefe modes of practice; and it would perhaps be ftrictly within the truth to add, their fatality alfo. On the firft eftablifhment of the purveying fyftem in the Bengal hofpitals, the enormous expenditure of wine, that favourite antifeptic, corroborant, cordial, and ftimulant, is alone fufficient to place. this affertion beyond a doubt.

Authors, but not of the firft eminence, have occafionally confined their practice to evacuations from the bowels in fevers of different defcriptions. Tiffot endeavours to enlighten the faculty with refpect to the treatment of certain fevers, by very confiderable evacuations downwards. It is, however, obfervable, that amongft many inftances of fuccefs attending thefe, the cafes of an unfortunate termination which he produces, though treated during their whole courfe by evacuations both ways, proved fatal after all, from a deficiency of thofe very means. Doctor Lyffons, at Bath, had practifed on principles of a fimilar tendency in fevers of different

deno-

denominations. Doctor Glafs, fenfible of the general prejudices againft purges to any extent in thefe difeafes, afferted their efficacy with great zeal. Sir William Fordice, whofe extenfive practice in London, during the greateft part of a long life, entitles him to every poffible degree of credit, has dedicated a confiderable treatife to the recommendation of inceffant purging in the putrid fevers of that place; in thofe of the inflammatory kind, Doctor Moore has teftified the utility of thefe difcharges. If a mind enlightened by original thinking, an intuitive fpirit of obfervation, and a fagacity which difcovers the fineft lineaments of difeafes, as well as of characters, demand our confidence, no opinion whatever fhould be allowed to ftand in competition with the fentiments of this gentleman. Doctor Duncan is an authority of great refpectability, in favour of purging in the nervous fever of children. We may omit others for the prefent; none of them, however, are to be found, if it is neceffary to except the latter, amongft the ingenious inventors of fyftems, or eminent profeffors of univerfities, who, unfortunately, in this inftance, are the authorities tnat influence the practice of moft gentlemen, until a larger or fmaller fhare

E 3                                    of

of experience, according to the proportion of underſtanding which each may poſſeſs, diſplays ſuch glaring truths, in direct oppoſition to the practice of thoſe great men, as neceſſarily forces conviction on their minds.

In the oppoſite ſcale, Doctor Cullen, and moſt of the profeſſors of that univerſity, Doctor George Fordice, and other lecturers in London, preponderate, as well from their number as from their celebrity. The generality of French phyſicians are more liberal of inteſtinal evacuations; but a true Britiſh ſpirit pervades the faculty, and compels them to view foreign practice with very little veneration.

It muſt be confeſſed, that Dr. Cullen's method of exhibiting antimonials in nauſeating doſes, to promote the cuticular ſecretions, is not only perfectly calculated to excite, but unavoidably productive of large diſcharges from the inteſtines; nor can they effect the former, exhibited in this manner, without a previous determination of the latter; ſcarcely any medicine operates with ſuch violence downwards, as nauſeating doſes of antimonials; yet the relaxation of the feveriſh ſpaſm on the ſkin is not, according to this gentleman of real genius, a ſecondary effect, immediately reſulting from the removal of the contents of the

the bowels, but a direct and primary effect of the medicine, independent of them, as the same cuticular spasm is suppofed to be the direct cause of the fever. With a perfect reliance on this hypothesis, as purging is thought to lessen the determination to the skin, he declares himself inimical to it beyond a degree that must be deemed in general very trifling.

As these pages are intended for the meridian of Bengal, no further notice need be taken of the variation which climate may cause in the nature and treatment of these disorders, than to observe, that the writer's individual experience, particularly during the passage of the Houghton, through every variety of climate, has given him the strongest suspicions, and perhaps no mean proofs, that the difference of the former is not so material as to require any peculiarity in the latter, unless in the degree which is admissible in the treatment of the same disorders in cold climates, according to their violence, and the various circumstances of age and situation in particular instances.

REMEDIES.—We now proceed to the consideration of the particular means usually employed in the treatment of fevers. It may first be premised, that from the nature of this paper

E 4                                                    it

it is neceffary to claim indulgence for an apparently prefumptuous rejection of the theories
of others, without the propofal of a better;
for fome obfcure hints without illuftration,
and many affertions without their concomi-
tant proofs; but it may again be repeated at
prefent, that there are ample materials to anfwer thefe objections, and to fupply thefe
deficiencies, to the fatisfaction of the public,
it is hoped, at no great diftance of time.

Diuretics may be neglected without any
interference with the prejudices, or rifk of the
difpleafure of the faculty; they now claim
little or no attention in the practice. It may
juft be obferved here, that the urine in fevers
does not exhibit any circumftances predictive
of the event, until other and more obvious
appearances have yielded lefs fallible fources
of conjecture.

A numerous clafs of antifpafmodics, corroborants, and cordials, may be treated with
fimilar neglect; fome of this clafs, however,
demand a degree of attention, which they
cannot receive at prefent beyond the very
narrow limits prefcribed to this paper.

Blifters do not poffefs, amongft the generality of practitioners in Bengal, that eminence which they have held in European
practice; even in that country they experience
undeferved

undeferved fupport : that they have any other
influence on the general fyftem, except what
proceeds from the pain they occafion, is very
doubtful; that their beneficial effects have at
any time exceeded the temporary fufpenfion,
or the entire removal of a morbid effect, or a
local action, is ftill more improbable; but
that they have ever contributed in any other
manner, either as antifpafmodic or ftimulant,
to the general removal of fever, appears
utterly impoflible, from a confideration of
the caufes affigned to fevers in Bengal, as well
as from experience. On the other hand, their
pernicious effects have been equally trifling,
unlefs where the mifapplication has occafioned pain, and a confequent increafe of the
febrile fymptoms. In this manner they have
often been the inftruments of much mifchief,
but have proved more frequently pernicious
by their infignificancy, and the blind confidence of the practitioner in fuch means, to the
neglect of others. Upon the whole, they may
be excluded entirely from the treatment of fevers with fafety, though their application may
fometimes procure eafe to a patient, and in
this way facilitate the removal of the caufe,
by the operation of other medicines, or contribute to the completion of a cure by their
action on local effects, during the exiftence

of

of the caufe, or on morbid habits after its removal.

The exhibition of bark on its firft introduction was confined to complete intermiffions; after a confiderable trial of its powers in that inftance, fome daring men ufhered it into fafhion during remiffions, when it foon procured admittance into continued fevers of every character, excepting thofe denominated inflammatory, and it reigns at prefent with a fway equal, if not fuperior, to antimonials. It muft be acknowledged, however, to the credit of the honourable company's medical fervants in Bengal, that it holds not fuch unbounded poffeffion of practice in its improved ftate in that country; yet the prejudices in its favour are ftill fuch, as to render it on many occafions an active poifon, to the deftruction of numbers. That it is a medicine occafionally applicable to the various modifications of fever, in the form of intermittent, remittent, and continued, as they occur in Europe, fhall not at prefent be denied; but that thofe occafions, even in the firft, are very unfrequent in warm climates, we may religioufly believe. During the firft ftages of every kind, before evacuations, powerful in proportion to the violence of the fymptoms, have taken place, its exhibition

bition is in direct contradiction to the nature
of the appearances, and to every indication of
cure, and confequently muft always be improper, and often fatal; but in no one
inftance, under any number and quality of
fymptoms, does its early exhibition appear fo
pregnant with deftruction, as during the firft
appearance of thofe fymptoms, which are
deemed highly putrid, and which, according
to the general opinion of the medical world,
demand its inftantaneous ufe in the largeft
poffible quantities. It is in vain to remon-
ftrate againft the abfurdity of attempting the
correction in lieu of the expulfion of putrid
matters. The patient's weaknefs, they will
affirm, evinces the impoffibility and the
fatality of the latter; and experience, univerfal experience, has proved the efficacy of
bark and wine. An obfcure individual op-
pofes fuch prejudices in vain; in vain he
afferts, that the general putrefcency cannot in
any inftance arife, unlefs from the putrid
contents of the ftomach and inteftines; that
fuppofing putridity to exift fometimes in the
general fyftem, it can only arife from a courfe
of abforption in the bowels; that, confequently,
fymptoms evincing the exiftence of thofe matters muft always precede general putrefcency;
that in reality the putrid appearances are in
general,

general, and always in the beginning, referable to this caufe, without any general taint;
that the moft obvious indication is the evacuation of thofe matters, where it can be effected
by any means the moft powerful; that the
general taint may invariably, in every inftance,
be prevented by early and vigorous evacuations; that the ftrength will ultimately be
found to increafe in proportion to thefe, until
the caufe be removed; and that then the
bark, in lieu of a poifon, may become of
fome ufe as a medicine.

Inftances, however, have occurred, fometimes in the moft violent cafes, when no
means whatever of procuring that evacuation
have been found to fucceed. From whatever
caufe this inert ftate of the bowels may be
fuppofed to proceed, bark and other powerful
medicines have either occafioned thofe evacuations, or have rendered the part fufceptible of the operation of evacuant medicines.
Thefe inftances, however, are very rare; they
feem occafionally to have arifen from the
ufe of large quantities of fpirituous liquors,
which, according to prevalent notions, fhould
have occafioned that inflammatory ftate of the
bowels which is deemed abfolutely prohibi-
tory of the bark.

What has been advanced with refpect to
the

the exhibition of this drug in putrid cafes
is applicable, in a certain degree, to all fevers,
to the nervous in particular, in which it can
feldom, if ever, be prefcribed with advantage, or without detriment. Intermittents
of the worft kind do not always require it,
and remittents ftill lefs frequently. It may
poffibly poffefs ftrengthened virtues, valuable
in proportion to its exemption from the heat-
ing qualities of other medicines of this clafs.
It frequently exerts a power by no means
equivocal, in ftopping the paroxyfm of a tertian; whether it be proper, or otherwife, to
accomplifh that object by fuch means is not the
queftion at prefent, and, perhaps, no medicine,
excepting opium, fo effectually operates to
the fuppreffion or deftruction of an effect,
after the removal of its caufe, or to the correction and prevention of the return of a difeafed action become habitual.

Notwithftanding the opinion has of late
years been nearly exploded, we muft forego
decifive experience, and renounce the evidence of facts, could we diveft ourfelves of a
firm belief, that the bark has often directly
or indirectly occafioned thofe obftructions,
and other inconveniences, in warm climates,
which on its firft introduction were afcribed
to its ufe.

Opium

Opium has been exhibited in various ſtages of different fevers. To aſſuage the violence, and to ſhorten the duration of a tertian paroxyſm, it was recommended with all the authority of Doctor Lind, whoſe opportunities of aſcertaining its effects were unlimited. But one might be led to ſuppoſe, that perſonal experience of its effects was by no means requiſite, previous to the liberal commendation of a medicine, ſince a late unfortunate gentleman of uncommon genius, but of a very inadequate ſhare of practice to enable him to aſcertain the virtues of medicines, ventured to extend the exhibition of opium to almoſt every ſtage of every fever. In particular circumſtances of fever, however, it has been noticed by ſeveral authors, particularly as of frequent utility in the nervous fever. So many more authors write *for* practice than *from* it, all affecting experience, that it is difficult to know the degree of credit which is due to their aſſertions. It has come within the ſcope of one individual's practice to afford its powers the faireſt poſſible trial in the tertians and other fevers of Bengal, as well as in the ſhip fever; and he can venture to give a pretty poſitive aſſurance, that its exhibition will, with the exception of very few inſtances, be found extremely miſchievous, even in combination with antimonials; this, however, will

not

not reft entirely on affurance, as the public fhall be foon in poffeffion of the proofs.

Of all articles of medicine or diet in dif-eafes, wine has been the greateft favourite, as equally fuitable to the prejudices of the prac-titioner and the palate of his patient. Phy-ficians in Europe, from choice or neceffity abftemious, have not acquired fuch an exten-five perfonal acquaintance with the virtues of this juice, as their brethren in thofe diftant regions; the fagacity of the latter has long deemed a liberal ufe neceffary, as a prophy-lactic, in the prefervation of their own health and fpirits in fuch oppreffive climes, in the counteraction of a relaxed fyftem of nerves, to which their companions are deplorably expofed, and in the recovery of their patients from difeafe; fo that in every predicament of life wine becomes a panacea. With thefe advantages it would have been very extra-ordinary indeed had the exhibition been con-fined to moderate potions. It required the hand of government in Bengal to reftrain the libations in the hofpitals within narrower limits. It would be invidious to dwell at prefent on the former exiftence of fuch a monftrous evil in the hofpitals of Bengal, fince it has already received a check from the fpirit (certainly an enlightened fpirit in this inftance at leaft) of public retrenchment. It

is

is probable the abufe of wine never did exift
of fuch a magnitude, in the honourable company's fhips; but as it may ftill be fuppofed
to exift in private practice to its utmoft extent, and, notwithftanding all reftrictions, to
a very confiderable degree in hofpitals, the
fubject cannot be difmiffed altogether without further notice.

In all fevers, in which appearances of debility prevail, without further reference to
the caufe, wine is prefcribed immediately,
confequently thofe fymptoms, which have
been called the nervous fever, have at all
times procured for the patient a large fupply;
but whenever putridity has been fuppofed to
characterize the fever, no quantity of this
antiputrefcent cordial, which the ftomach
could retain, has been deemed exceffive; accordingly the patient has been gorged, until
the violence ufed has, in fome inftances,
forced the moft copious and moft falutary
evacuations by vomit or ftool, but in general
an increafe of coftivenefs, fever, delirium, &c.
has enfued. If the abufe be fuch during the
prevalence of fymptoms of this character,
what muft be the magnitude of the error,
when it is exhibited during any period, but
particularly at the commencement of fevers,
which the practitioner himfelf calls bilious,
though fome doubtful fymptoms of pu-

5                                    ⌊trefcency

trefcency fhould be thought to accompany them?

Experience has afforded the fulleft conviction, that wine is very ill adapted to any period of any fever in thofe latitudes, when the treatment has been proper in other refpects. The general exhibition of a medicine, whofe merits depend on the mifapplication of other means, is no evidence of an improved ftate of practice. While the character of a medicine faves the practitioner the trouble of obferving and prefcribing after nature, it poffeffes no fuch faving powers for the patient. Were the virtues of this article on any occafion, or in the fmalleft quantity, effential to the cure, or to a recovery from a ftate of convalefcency, the native foldiers of Bengal would have very little chance of either; for their religious prejudices compel them very generally to confider wine with fentiments of the utmoft abhorrence, even under the palliating appellation of a medicinal mixture. Of the unconquerable obftinacy of this prejudice ample proofs were afforded on the writer's firft arrival in India, in hofpitals containing nearly three hundred natives, at a time when he had not as yet divefted himfelf of the prepoffeffion in favour of wine, which he had imbibed in Europe; at this period, confe-

F                    quently,

quently, he prefcribed it on all occafions with great liberality.

It will probably be found a vulgar and pernicious error, that perfons who have been accuftomed to the free ufe of wine during health, require a greater proportion under the preffure of ficknefs; the very reverfe may poffibly admit of proof.

He may venture to declare, that in a wide field of inftruction from cafes, he has not obferved one, in which he could pronounce the exhibition of wine as a medicine at all eligible, and not many which appeared to require it as an article of diet; but that he has witneffed innumerable inftances of its dangerous and fatal effects in fevers. No doubt, the mifchief occafioned by wines is in proportion to the quantity of fpirit in each; the weakeft, confequently, will prove the moft innocent. Claret may poffibly have the advantage of port and madeira, and hock may be ftill lefs detrimental than the firft. Wine of every kind would feem to deferve a preference to bark and opium, as it is not attended with effects equally permanent in fuppreffing the fecretions of the bowels. Some wines are even thought to poffefs a laxative power; bark alfo, when it operates by ftool, as it fometimes appears to do with

confi-

confiderable activity, cannot perhaps be ac-
cufed, with juftice, as the author of much
mifchief.

These obfervations on bark, opium, and
wine, are indeed very repugnant to the doctrines of the late Doctor Brown; but it is
hoped, that the number of his difciples, or
the prevalence of his methods of treatment,
is entirely confined to thofe gentlemen, who,
like Doctor Brown, have had little opportunity of exercifing the practice on any but
themfelves. To fuch, however, whofe minds
are more fufceptible of the impreffions of
novelty, without adequate proofs, the moft
pofitive affurances may be given of the dangerous confequences that would inevitably
refult from fuch a liberal and indifcriminate
ufe of thofe powerful medicines in the fevers
of hot climates. Doctor Brown's practice
may poffibly apply to fome circumftances of
low fever in Europe; but the occurrence of
fuch fevers will be found very rare in the
meridian of Bengal, particularly in the upper
provinces, where moft acute diforders have,
what would be deemed in Europe, and according to Doctor Mofely, in the Weft Indies,
an inflammatory afpect, and are alfo characterifed by larger fecretions of bile. Whether
thefe fecretions be deemed a caufe or a

confe-

confequence of the diforder, they will furely
be allowed to require active means for their
immediate difcharge. We may fubmit to
the determination of any confiderate practitioner, the merits of Doctor Brown's fyftem,
under fuch circumftances as thefe. This
gentleman's tenets demanded particular notice here, as feveral addreffes, attributed to
furgeons of the honourable company's fhips,
and recommendatory of Doctor Brown's practice, were publifhed in the news-papers at
Calcutta, during the years 1788 and 1789.

Doctor Robertfon's late publications furnifh a recent and broad contradiction to the
fentiments expreffed in this paper on the fubject of thefe medicines, particularly of bark.
With what degree of confidence can a tyro
in the practice of phyfic perufe fuch a variety of contradictory methods of treating the
fame difeafes? Is it not extraordinary that
patients, who are perfectly aware of this dif-
traction of medical opinions influencing practice, fhould place any dependence whatever
on the efficacy of the fcience, or the fkill of its
votaries? Doctor Robertfon's extenfive opportunities give him a juft claim to pronounce a pretty pofitive judgment on this
fubject; others arrogate a title of equal refpectability in favour of their peculiarities.
Expe-

Experienced phyſicians, as well as the
youngeſt members of the faculty, muſt be
confounded with this great variety of con-
tending atoms in phyſic. What opinion
can poſſibly be entertained of this chaos ? I
can only aver, that it has contributed to render
me extremely diffident of the accuracy of my
own ſenſes, and of the evidence of the moſt
ſtubborn facts. It is in vain to ſay, "Me-
" dio tutiſſimus ibis;" that the point of per-
fection, or of ſafety, is exactly in the middle of
this circle of extremes. From the days of Hip-
pocrates to this hour, that imaginary point
has not been aſcertained; the advice has been
often repeated, the declaration often made,
but promptitude of aſſertion, and facility of
proof, are often entire ſtrangers.

Is a fever, like a nuiſance in the ſtreet,
which may be equally well removed in
every oppoſite direction of the compaſs, re-
moveable from the conſtitution by a contra-
riety of means ? Phyſicians ſeem to think,
that it can make its exit in every direction,
through the ſkin and other outlets, whoſe
ſites are the reverſe of one another. Will
the means of expulſion be allowed to be as
various and oppoſite as the channels ? This
view of the ſubject is by no means ludicrous.
Our knowledge of firſt cauſes and effects in

F 3                    medi-

medicine is fo limited, that we cannot deny with propriety, but what remedies of an oppofite operation may have the fame ultimate effect in the reftoration of the conftitution to its natural habits; and perhaps the ways which lead to this object may be as various as the high roads to a metropolis.

A conjecture of this extraordinary nature can alone reconcile the contradictory merits of Doctor Cullen's, Doctor Brown's, and other methods of treating fevers. But we may be more rationally inclined to think, from this mafs of pofitive affertion, and apparent and plaufible, if not certain proof, that it is extremely difficult to kill in fevers; and that, when death is not the confequence of the patient's own efforts, the practitioner muft be endowed with the moft uncommon powers of deftruction to accomplifh this purpofe.

VENESECTION.—Perhaps a general divifion of remedies into fuch as promote and oppofe evacuation, would form a tolerably accurate outline of diftinction; moft of thofe which have been noticed are of the latter clafs; and we may enter on the fubject of the former with the article of venefection.

Symptoms,

Symptoms, which attend the commencement of fevers in Bengal, will appear on a
comparifon with the fymptoms of inflammatory fever, in many inftances exactly fimilar; yet the inflammatory fever, independent
of local affections, is generally acknowledged
not to have exiftence either in fhips or on
fhore in that country. The full, ftrong,
rapid pulfe, red eyes, bloated and florid countenance, burning fkin, high-coloured urine,
parched tongue, &c. are frequent attendants,
particularly on the firft attack of fevers,
which occur chiefly in the feafon of the hot
winds, and in conftitutions habituated to exceffes. In thefe fevers the ufual effects of
venefection appear at firft very favourable, by
a total remiffion of the exceffive heat, through
the means of a profufe fweat; if fortunately of
an inteftinal evacuation, which fometimes occurs, the benefit may be permanent, though
the patient is unneceffarily weakened. It is not
often, however, that the inteftinal difcharge,
in this inftance, proves permanently beneficial. The practitioner's mind, rivetted on
other means, overlooks the benefit occafioned by this difcharge, or attributes it entirely to the bleeding; and although he may
not be intrepid enough to repeat this, yet his
whole attention is directed to the fupport of

F 4                                               its

its effects by perfpiration, which he deems
the natural and happy crifis of thefe diforders. The fweat which flows after venefection is very profufe, and of courfe debilitating. This difcharge is fupported with
unremitting affiduity, to the great diminution of the patient's ftrength, and the increafe
of an obftinate coftivenefs. Tyfot, Doctor
Paifly, and others, have obferved, that venefection in certain fevers occafioned delirium
at fome future period of the complaint. Inftances have occurred, in which delirium did
really appear to follow the venefection. The
deceitful interval of eafe, which fucceeds this
operation, foon vanifhes; the fever rekindles,
probably with redoubled fury, and finds the
patient much lefs capable of refiftance to its
violence, exhaufted by the bleeding, and by
the exceffive difcharges from the fkin, which
have been imprudently prolonged by various
means.

Thefe fevers frequently commence with
every fymptom of a high degree of inflammation in the region of the liver, which, as
they arife from the foul contents of the
ftomach, duodenum, &c. fubfide immediately
on the evacuation of thefe. It is obvious,
that the ufe of the lancet on fuch occafions
is unneceffary, if not pernicious; yet it often
proves

proves the firft refource of medical gentlemen on their arrival, as well as of thofe who
have had more opportunities of obferving,
and have acquired a greater fhare of experience. Many cafes of real inflammation in
the fubftance, membranes, or appendages of
the liver, are attended by fymptoms of confiderable fever; but on thefe occafions the fever
will feldom be perceived to precede the local
affection, as in the former, and the pulfe will
fuggeft an idea of hardnefs, not equally
ftriking in the other. The difcrimination is
not very difficult to an experienced perfon,
and cannot be difcuffed to a greater extent in
this paper. The operation of purgative medicines will afford the beft proofs of the
nature of the complaint, and may, in all
doubtful inftances, precede venefection with
fafety and propriety.

When fymptoms of incipient putrid fever
occur, the cautions againft this remedy are fo
pointed and numerous in authors, that few
gentlemen will venture to prefcribe it; but it
is not always a very eafy tafk to the inex-
perienced to form a decifive opinion on the
nature of the fymptoms under their infpec-
tion. Inftances of the ufe of the lancet, in
almoft every variety of fever, have occafionally come within obfervation. It has not
only

only been known in acute cafes, but alfo in
chronic fevers of great obftinacy, in which it
has been ufed by gentlemen of real experience, when they have appeared totally at
a lofs how to proceed, after the failure of
other means; but on no occafion, to the
beft of my judgment, has this remedy been
attended with ultimate fuccefs. Fevers fometimes occur in thofe latitudes, which, after
one or two regular returns of cold and hot
ftages, affume the continued form, and pre-
ferve it very obftinately, notwithftanding inteftinal evacuations, with fymptoms nearly
refembling the inflammatory. On fuch occafions, venefection will, no doubt, fometimes procure an immediate remiffion, but
with a rifk of a fatal termination on a return
of the fymptoms.

This remedy is never fo likely to be mif-
applied, as in puerperal fever, on the inflammatory or putrid nature of which the moft
refpectable authors are divided. It has probably often proved fatal in this diforder,
fometimes of uncertain effect, but very feldom advantageous, where no obvious local
affection has prevailed. It may be affirmed,
that as even this fever, in warm climates,
originates from a fource which venefection
cannot affect, the vitiated contents of the
ftomach

ftomach and inteftines, its ufe fhould be exploded, or greatly reftricted, in the puerperal
as well as in all other fevers of Bengal.
There is fome reafon, however, to fear that
the influence of an authority, fuch as Doctor
Mofely's, might tend to introduce into Bengal practice a remedy which he has declared
himfelf to have found extremely fuccefsful in
the Weft India difeafe, which he calls the cau-
fus, or ardent fever. The fame form of fever is
by no means unfrequent in Bengal; but it is
to be hoped, that the fame remedy will always
remain of very unfrequent ufe in that country.

The violent affections of the head, cheft,
and other parts, which frequently attend
fevers, feem to indicate the inftantaneous ufe
of the lancet; but as thefe arife from a fimilar
fource, they readily yield to fimilar means.

SWEATING.—The excitement of fweating, at the commencement of fevers, by violent means, with few exceptions, is now uni-
verfally exploded. But the practice of raifing
and prolonging the cuticular fecretions by
remedies, which are gentle and gradual in
the production of their effects, is adopted
as generally. Antimonials are made principally fubfervient to this purpofe; and notwithftanding the erroneous, though prevalent
mode

mode of reafoning on their effects, it will
readily be allowed, that they are perfectly
calculated to anfwer this indication, by their
powerful and fometimes violent action on
the ftomach and inteftines, the confequent
removal of vitiated contents, and ultimate
reftoration of the cuticular difcharges. But
it is equally evident that, if this be the progrefs of their operative powers, the affection
of the fkin, under its various and favourite
names of fpafm, &c. &c. will be degraded
from the rank of a powerful proximate caufe
to that of a fecondary effect; from the proud
eminence of the principal object of a phyfician's attention, to total or nearly total neglect;
and that fweating will fometimes even require
a ftrenuous counteraction; that, as it does not
appear confiftent with reafon to infift on the
exit of an unwieldy mafs of corrupted matters, by the labyrinths of the circulation from
the bowels through the fkin, fweating, though
in fome inftances it may afford a temporary
palliation, at the expence of ftrength, cannot
remove the caufe of the diforder, and muft
confequently prove often ufelefs, in fome cafes
egregioufly trifling, and in many others dangerous and fatal. What authors are pleafed
to call the *low nervous fever* forms no exception to this decifion. As the diaphoretics
ufed

ufed in this diforder exert their activity more on the general fyltem, and lefs on the particular fecretions of the bowels, they become, proportionally, more deftructive.

To avoid prolixity at prefent, a chain of affertions may be offered, of which the connection and validity fhall appear in a future publication. He may therefore affirm, that the foul contents of the ftomach or inteftines are capable of exciting the cuticular fecretions morbidly; that in the inftances of fpontaneous fweats defcribed by authors, which have not proved beneficial, thefe have probably been the exciting caufes, and, in conjunction with mal-practice, have had a fhare in all forced fweats; that from the entire expulfion of thefe from the body, or their removal from the neighbourhood of the ftomach, all fweats of a favourable termination, and fuch as have been efteemed critical, have flowed; that fweats afford not any conjecture refpecting the event, which may not previoufly be formed from an obfervation of the progrefs, or the actual predicament of the exciting caufes in the bowels; that, in the increafe and diminution of the oppreffion about the præcordia, and other fymptoms, which appear to accompany fweating, authors have uniformly miftaken the caufe for the
effect,

effect, and the effect for the caufe; that the
moft powerful means to promote perfpiration
confift in the removal of the caufe of the
fever, by evacuating the foul contents of the
ftomach or inteftines, principally by purging;
that as an effect may continue to operate for
fome time after the removal of its caufe, or a
difeafed action may become habitual, perfpiration will not always fucceed immediately
to the expulfion of the offending matters, nor
the fever ceafe; that in a protraction of this
nature, medicines which do not evacuate the
bowels, may be poffibly adapted to the counteraction of thofe effects, and the reftoration
of natural habits to the fkin; that all diaphoretics, not direct evacuants, whenever they
have proved ferviceable, have operated in
this way; that on thefe principles, attempts
to excite perfpiration in any other way, during
any period of fever, than fuch as promotes
evacuation from the ftomach and inteftines,
are always to be avoided, and a fpontaneous
tendency that way to be checked by cool air
and other means, until evacuations have
taken place; and that, confequently, almoft
the whole tribe of fudorifics fhould be exploded from practice in Bengal.

VOMITING.—Perhaps there does not exift
a fingle fpecies of fever, in which vomiting is
omitted

omitted by practitioners and authors of every clafs. It is generally, indeed, the very firft refource, and in reality appears, at firft fight, the moft eligible and effectual means of expelling the matters which excite fever. Antimonials are, or ought to be, the only medicines employed for this purpofe. Thefe are commonly adminiftered in fuch dofes as produce and fupport a tendency to vomit for fome time before the actual evacuation takes place. This is the form of exhibition moft approved by the whole body of the faculty, but particularly by the Cullenian fchool; and it is prefumed that the honourable company's medical fervants, as well in their India fettlements as on board their fhips, have, with very few exceptions, conformed at laft to this fafe, and often efficacious practice. We may pafs over the doctrines, which attribute the immediate effects of emetics to antifpafmodic, mechanical, or other virtues, and in few inftances to their evacuating power, either directly affecting the ftomach, or, ultimately, the inteftines, and promoting frefh and healthy fecretions from all the bowels; and we may affert, in general, that it is by the exertion of thefe primary properties alone they prove fometimes of fuch extraordinary efficacy in fevers. In a former article we have anticipated

pated the nature of their operation as fudorifics; and we may now venture to fuppofe,
that they poffefs not any fpecific charm to
fafcinate and deftroy fever, although they
muft be acknowledged well adapted, powerful, and fuccefsful inftruments of health, in
the hands of perfons even of common fkill
in thefe diforders; yet frequent trials have
long afforded a conviction, that their exhibition is generally premature, fometimes altogether unneceffary, and not unfrequently attended with danger; that their ufe fhould
not precede evacuations of the groffer matters, lodged as well in the lower inteftines as
in the vicinity of the ftomach, unlefs in cafes
of impending fate, when the former cannot
be fpeedily accomplifhed, or when fpontaneous vomiting and violent retching render
their exhibition eligible; on thefe occafions,
even the antimonials fhould be joined with
fome purgative medicines, principally in a
liquid vehicle, to give the matters a final
tendency downwards, through their natural
channel; that the action of vomiting, before
the neighbourhood of the ftomach and the
inteftines have been in fome degree unloaded
by purgatives, may force into the ftomach
matters more pernicious in quantity as well
as quality, than thofe which had a previous
exiftence

exiftence there; that; confequently, the diftrefs and danger may be aggravated by fuch
means; that inteftinal evacuations will frequently fuperfede the neceffity of ufing
antimonials in that way; that, in many
inftances, at the beginning of violent fevers,
antimonials, particularly in the naufeating
form of exhibition, prove too feeble, and
ultimately fatal, from the blind confidence
repofed in them, to the exclufion of means
as powerful and rapid in their operation as
the fever they are intended to oppofe; and
that, at certain dangerous periods of fever,
when nothing but the moft active purgatives
will fave the patient, fubdue the diforder, and
reftore ftrength, antimonials in any form, or
emetics of any kind, even when they operate
in the moft favourable manner, as inteftinal
evacuants, are generally deficient in the vigour of this very operation, while they contribute, in a more particular manner than
mere purgatives, to deftroy the remaining
ftrength of the patient.

The action of the ftomach, which accompanies naufea, is admirably calculated to promote the difcharge of its contents into the
inteftines, and fo far may prove the very beft
means of cure. But this action gives a great
fenfe of debility to the ftomach, as well as to

the fyftem at large, and medicines merely
purgative do exift, which anfwer the fame
purpofe, unattended by fimilar inconveniences. It is acknowledged, however, that
in many inftances their affiftance becomes
indifpenfible, in flighter attacks perhaps
adequate to the total removal of the diftemper, and always neceffary at fome period or
other of fevers of confiderable duration.

Although the writer difclaims every wifh
to depreciate the value of thefe medicines,
either in the form of active emetics, or of
naufeating dofes, while he endeavours to procure precedence to more natural and more
powerful means, yet the apparent difrefpect,
which he may be thought to have fhewn to
thefe favourite indications of cure, may poffibly excite the reprobation of very enlightened and experienced practitioners, even in
Bengal; he fhall truft to a very early publication of the proofs, and to a more general
diffufion of the practice recommended in this
paper, for his complete juftification.

PURGING, or, THE CURE.—As the method
of treating fevers depends principally, if not
entirely, on the ufe of proper purgatives, it
will naturally be contained in this article.
The confideration of glyfters may with propriety

priety have place here; but this paper has already fo far exceeded the limits which were at firft prefcribed to it, that the whole of this important article of purging muft be treated with much more concifenefs than could have been wifhed or intended. We may only ob-ferve, with refpect to glyfters, that they can feldom be expected to evacuate more than the contents of the rectum; that by their affift-ance in this particular they facilitate the operation of purgative medicines, which they may always be allowed to precede or accom-pany, until the groffer matters have been dif-charged; that, after this period, though their operation is feldom productive, yet a languor and degree of weaknefs will generally be found to enfue; that their virtues, during a ftate of convalefcence, are too trifling to remove or to obviate coftivenefs, and prevent a relapfe; that their adminiftration is allow-able only at the commencement of fevers, unlefs the vicinity of the rectum be loaded with groffer fœces during any other period; but that the reliance, which practitioners have mifplaced on their evacuating powers, has often rendered them inftruments of de-ftruction, notwithftanding their general in-fignificancy in thefe complaints.

We may now proceed to the principal ob-

ject

ject of thefe fheets, the treatment of fevers
by active purges.  In confirmation of the validity of the writer's remarks on other medical means, the proofs are chiefly negative, as
he has not often experienced the neceffity of
their application; but fuch as he fhall ad-
duce on a future occafion, in fupport of the
indifpenfible neceffity and extenfive ufe of
purgatives, are of fuch a pofitive nature,
evinced in fuch a variety of inftances, and
corroborated by fuch ftriking contrafts in the
modes of treatment, that they cannot fail to
force conviction on the moft bigotted minds.
At prefent, however, he fuft confine himfelf
to general affertions.

It has already been obferved, that the
practice of gentlemen in Bengal, at leaft in
fome fevers, is deficient rather in the extent
than in the nature of the means.  This reflection, however, may be reftricted to the
moft experienced.  Thofe who truft to information acquired from books and lectures
will naturally crawl on in a blind routine,
proceed greater lengths in error, and mifap-
ply means altogether.  The difference between thefe two claffes confifts in the gentle
exhibition of inteftinal evacuants, and in the
total, or nearly total, omiffion of them.  The
former have added to their own experience

ence the authority of authors, to whofe writings the latter have not had accefs, or whofe opinions, unfupported by a glare of reputation, they have defpifed.

. From the earlieft periods, decided opinions in favour of the exhibition of purgatives may be detected in authors of every clafs, as far as opportunities of confulting them have offered. Some amongft the lefs illuftrious of modern writers, it has been already mentioned, recommend them to a confiderable extent; but not one, as far as reading ferves on this fubject, to the degree and in the form which becomes indifpenfibly neceffary in moft inftances.

The neceffity of purging is not reftricted to any fpecies of fever, but affects every fet of fymptoms ufually denominated fever, in proportion to the multiplicity and dangerous tendency of thofe fymptoms. In that country, many modifications of fymptoms, fuppofed to conftitute diftinct fevers, particularly the putrid nervous, derive their very exiftence from the neglect of thofe means at an early period. We may fufpect, that what is generally underftood by the appellation of putrid fever, under the various forms and circumftances of fhip, hofpital, jail, puerperal, and in Bengal, jungle, pucka, and hill fever, would never have exifted the

terror

terror of mankind, had nature been diligently
watched, and thofe means of prevention ufed.
By the acknowledgment of all authors, putrid
fever feldom affumes its characteriftic fymptoms on the firft notice of the indifpofition,
In fituations and circumftances of thofe climates, in which fevers of every kind might
be deemed moft liable to degenerate into that
form, fuch putrid appearances have not been
known of any degree of violence, during the
fuccefsful treatment of fome hundred cafes.

Purging, therefore, on the firft hints from
nature, will generally obviate the accefs of all
fevers in every predicament. On the firft
attack of thefe, purging will infallibly prevent the approach of dangerous fymptoms,
particularly thofe called putrid, and, at their
height, will always fave, and generally cure,
the patient. It is not to be expected, that
in every inftance they fhall anfwer this character without fome affiftance from other remedies, particularly emetics. Cafes have occurred, in which the immediate and rapid
exertion of every power, productive of vomiting as well as purging in their utmoft vigour,
feemed fcarcely fufficient to refcue the patient
from almoft certain death; and it was deemed
very fortunate when thefe effects could be
produced by any means the moft powerful,

8

In

In flight cafes, the quality as well as the quantity of the purgatives in common ufe may fuffice. The faline and the oily, in whatever vehicle, do not appear to operate much beyond the expulfion of the groffer contents of the inteftines. The former very feldom contribute to convey any impurities of the ftomach into the inteftines; and their operation, after the firft difcharges, is confiderably debilitating.. The action of the latter, it muft be acknowledged, is often productive of large bilious difcharges, and their efficacy, after the firft evacuations, fuperior to the former. Caftor oil does not appear to increafe the feverifh heat as much as the others; nor have thofe pernicious qualities, which are attributed to oily medicines in fevers, been perceived. That they may not, however, remain too long in the bowels, their operation fhould be rendered certain and early by a repetition. This is a precaution that fhould never be neglected after the exhibition of any purgatives; the action alfo of this oily medicine may be pretty confiderable in the ftomach. The peculiarities of thefe claffes of purgatives may probably render each applicable to particular periods and circumftances of the difeafe. The faline purgatives poffefs one very great advantage over the other

G 4                                  kind;

kind; their activity may be greatly increafed by the addition of various proportions of tartar emetic. By this combination, their action on the ftomach and the fecretions at large prove very confiderable. The faline and oily combine with advantage.

Very few of the other orders of purgatives deferve any attention in thefe diforders, or perhaps in any others in that country. The fymptoms fometimes yield altogether, or abate confiderably, after the operation of thefe medicines; in all cafes they fhould certainly be allowed precedence. But as the difeafe does not always arife from the quantity or quality of groffer matters in the ftomach and inteftines, or from any proportion of vitiated bile and other fecretions, which the utmoft power of thefe purgatives can affect, we muft have recourfe to fuch as are more active and better calculated to remove the caufe of the complaint, which may frequently be fuppofed to arife from the quantity, deficiency, vitiation, or immobility of certain fecretions of the ftomach and inteftines, particularly the mucus adhering to the latter. That the caufe of the protraction of fevers is often connected with the ftate of the mucus, as well as of the other fecretions, appears from the immediate ceffation or alleviation of all the fymptoms

toms on a copious difcharge; and that the
mucus is often vitiated in a moft extraordinary manner, the fenfes of the obferver will
afford ample teftimony. There are practitioners, to whom thefe cannot prove a fource
of information. The extreme delicacy of
fome gentlemen will not permit them to
carry their refearches fo far ; yet it is from
this fource, and this alone, that any precife
knowledge refpecting the nature, probable
duration, and other circumftances of the diforder, but particularly the neceffity of further evacuations, can poffibly be acquired.
It may be deemed particularly fortunate, that
the purgatives which prove moft fuccefsful
in fevers are as mild in their operation as
they are certain and powerful; that they are
not fubject to the inconveniences attending
the other claffes, for from their want of
bulk they are more retainable in the ftomach ; and that from their fpecific gravity
they may be fuppofed to reach more readily
the fources of the evil, and to combat thefe
with more fuccefs. Mercurial purgatives,
particularly calomel, poffefs thefe advantages
in the trifling quantity of two or three
grains; but fuch fmall dofes are feldom of
much efficacy after the firft and fecond, and
a repetition would be efteemed rafh by the
generality

generality of practitioners. They have frequently, however, in the smallest proportion,
an operation so extensive, as to remove the
complaint altogether, in slighter cases, by
copious evacuations. But other occasions
require their exhibition in such quantities,
and after intervals so short, as would terrify
most of the faculty, even in India, and appear
to practitioners in Europe necessarily fatal.
The most trifling detriment, however, has
not been observed in any one instance, though
a discharge from the salivary glands has not
unfrequently ensued. It is always, however,
proper, as well to obviate these inconveni-
encies, as to render their evacuating powers
more certain, to urge their operation by
other cathartics, especially in a liquid form.
It may be received as a general rule, that the
calomel, either alone or in conjunction with
cathartic extract, resin, or extract of jalap,
scammony, gamboge, elaterium, or the mass
of laxative mercurial pills, should be exhibited
at night, and the medicines necessary to promote its effects early the ensuing morning, as
well as during the course of that day, according to circumstances. From two to
ten or more grains of calomel, with a greater
proportion of any of the other articles, may
form a dose with the utmost safety; for these
medicines,

medicines, as evacuants, do not act with a
difturbance, nor perhaps with an efficacy, in
the exact proportion of their quantities.
Thefe dofes may and fhould be repeated every
fecond night, or, according to the preffure of
the fymptoms, every night, as long as any
thing offenfive fhall remain to be difcharged
from the bowels, in the form of groffer ex-
crement, vitiated bile, mucus, &c. Forty or
more grains of calomel, with a larger quantity of the laxative mercurial pill, have been
exhibited with innocency, and with great benefit, in this manner, during the courfe of
five or fix days. Laxatives alone, or with
additional efficacy from an union with antimonials, fhould be adminiftered, not only in
the mornings after the calomel, but in fmaller
quantities during the whole of the intervals;
a very dilute folution of tartar emetic alone
generally anfwers this purpofe extremely
well.

As fymptoms called putrid, nervous, &c.
indicate the excefs in quantity and vitiation
of the offending matters, and confequently
the greater obftinacy and danger of the diforder, notwithftanding the general prejudices
againft the ufe of mercurials in putrid cafes,
this courfe of purging by calomel is more effentially neceffary when fuch fymptoms prevail,

vail, than on any other occasion whatever.

. There is little doubt but that puerperal
fever, which a celebrated professor of mid-
wifery in the university of Edinburgh deemed
absolutely incurable, may be prevented invariably by effectual evacuations from the intestines after delivery; when it does super-
vene, it will be found to admit of a pretty
certain cure by purgatives, given in quantities proportioned to the violence of the
symptoms, with very little attention to the
apparent weakness of the patient.

Fevers which resist this mode of treatment
when they have been neglected or ill treated
for some time, and degenerate into, or have
originally assumed the different forms of slow
fever, whether the cause be obstruction or
otherwise, will generally yield to mercury in
its various preparations, exhibited in frequent and small doses with occasional purges.
Mercurial inunction has been attended with
very general success, not only in these slow
chronic fevers, but also in the violent, acute,
burning fever, which has been denominated
by a variety of appellations in the West
Indies, such as yellow fever, black vomit,
&c. as particular symptoms seemed most prevalent to each practitioner. This fever, with
every

every fymptom by which Doctor Lind and
other authors have characterifed it, has occurred in Bengal. In fome unfortunate cafes,
the differences exhibited proofs of a violent
affection in the liver and fpleen. After this
difcovery, mercury, exhibited fo as to affect
the mouth as foon as poffible, with occafional
laxatives, proved uniformly fuccefsful. In
this inftance, the difeafe is fo rapid in its
progrefs to deftruction, that the exhibition of
mercurials fhould be equally rapid and vigorous. But the happy effects of a more
gradual courfe of mercury are juft as ftriking
in thofe flow fevers, which would be called
hectic, nervous, &c. by European phyficians,
but which in that country fhould be efteemed
the confequence of neglect or mal-treatment
of preceding fevers in the continued, remittent, or intermittent forms. It will generally be found that thefe chronic fevers, whether they afflict the conftitution without any
fenfible periods of abfence, or only return in
occafional relapfes of more feverity, will yield
equally to the operation of mercury on the
fecretions, with the intervention of purgatives; they will alfo be found, during their
firft attack, to have refifted the utmoft efficacy of the bark, and other medicines in
common ufe. It may be neceffary, however,

to

to fuggeft a caution to the practitioner, that
he should not think himself difappointed, if
the operation of mercurials do not always
appear to be attended with decifive effects,
though the falivary glands should be confiderably affected; for the ultimate benefit
from this courfe may not be very evident for
one or two months after its ceffation; at
laft, however, returning health and embonpoint will convince the practitioner of the
fuccefs of his efforts. During the treatment,
the reftoration of the fecretions of the bowels, particularly of the liver, is fometimes
attended with fuch apparently difagreeable
fymptoms, that the practitioner may be led
to form an unfavourable judgment of the
plan. The formation or the difcharge of
·bile, which has been fuppreffed, deficient, or
irregular for a long time, will not unfrequently be attended with fevere fymptoms of
dyfentery and fever. To a perfon of experience thefe will afford the moft favourable
omen of the ultimate fuccefs of his remedies.
Nothing, however, affifts the falutary agency
of mercury with fuch power, as frequent
changés of fituation and air, with a diet of
mild vegetables and water only, and frequent
exercife of the gentler kind. In reality, a
fea voyage, with a ftrict adherence to this
diet,

diet, would alone prove fufficient, in moft inftances, for the entire recovery of the patient, and would confequently fuperfede the neceffity of mercurials.

Thefe are the fevers in which the lunar influence is moft obfervable; particularly when they are accompanied by any local complaint. The operation, however, of this influence is by no means equally remarkable through all feafons of the year. The unhealthy feafons are of courfe thofe in which it is moft apparent; and it has already been noticed, that the feafon of the rains, or the termination of the periodical rains, is particularly favourable to the production of fevers, fluxes, and other diforders. This influence is confequently very evident at thofe periods, and fuggefts the neceffity of precautions to counteract it. The principal prophylactic is inteftinal evacuations before the different periods of the moon. In recent fevers, it is imagined, this influence will be very obfcure at all feafons of the year; nor is any particular attention to it neceffary during the cure, In fevers of great length, which never occur, perhaps, but from mifmanagement, it might poffibly be improper to treat it with entire neglect. But what will the moft fcrupulous obfervance of thofe periods avail, if the treat-

ment

ment be in other refpects erroneous ?—As the merits of this long-depending caufe will be allowed to fpeak for themfelves in a body of evidence which is now in the prefs, with an entire indifference to the opinion of the public, and perfect exemption from all pre-judice on this fubject, no attempt fhall be made to influence the minds of others, but the ultimate decifion fhall be abandoned to fuch readers, as may efteem the fubject of fufficient importance to deferve a particular fcrutiny.

In all varieties of fever, no doubt, other remedies, of a nature very different from evacuants, may occafionally have place du-ring thefe courfes. But we cannot enter into a minute difcuffion of every part of the treatment in this paper, the principal object of which is to inculcate the indifpenfible neceffity of inteftinal evacuations by any means, but efpecially by large dofes of ca-lomel.

With refpect to diet, we may only obferve, that it is not, in general, fufficiently re-ftricted; and that fifh of every kind, as well as all animal food, to the utter exclufion of chicken broth, fhould be prohibited during the exiftence of any lurking fymptom of fever, however trifling it may appear. Even
freſh

frefh vegetables are frequently inapplicable to the purpofes of diet in fevers; fo that farinaceous aliment alone, or with mild and ripe fruits, would feem to be the beft adapted to every period of thefe diforders.

It will fcarcely be underftcod, that the whole of the practice recommended in the foregoing pages fhould be extended to patients reduced by ill treatment or neglect to the laft degree of emaciation and debility.

As it will appear in a very fhort time, that the obfervations contained under the article of fever, are the refult of a fuccefsful mode of practice, what fentence fhall we be compelled to pafs on doctrines and practices nearly the reverfe, though promulgated by men efteemed the greateft lights, but in this inftance the " ignes fatui," of their profeffion, when thofe opinions are deemed applicable to the treatment of fevers in Bengal? What can we fay refpecting authors, who have written on fevers in climates and fituations fomewhat fimilar to the one which has juft been mentioned? Shall we permit our judgment to be blinded, our experience to be entirely difcredited, and our lives to be facrificed to the fame which blazons the volumes of feveral authors, whofe treatifes are in every gentleman's poffeffion, and whofe practice, of

H courfe,

courfe, prevails as generally as their names
are known? Some have written on the difeafes of thofe latitudes, who have never ap-
proached them, except in contemplation.
Others have deemed the refidence of a month
or two on the fpot fufficient to give a ftamp
of authority to their writings. Their ingenuity has fafcinated the medical world in
thofe parts, to the deftruction of thoufands.
Shall we treat fuch abilities with refpect, or
even with lenity? or fhall we venture to oppofe the evil by a rafhnefs which will be
deemed very extraordinary, and to declare,
that thofe who poffefs the beft claims to our
attention, by a temporary refidence in the
Eaft Indies, do not appear to deferve credit
for more than a good intention, induftry, and
ftone-blind experience? May we not allow a
very fmall fhare of judgment and fagacity to
thofe, who have appeared uniformly to miftake the falutary efforts of nature for the
morbid effects of the difeafe; to oppofe thofe
efforts with the moft fatal pertinacity; and,
finally, to recommend a practice, the mortal
tendency of which will appear from the
broad evidence of their own writings?

When the obfervation and experience of
fuch men are arraigned, we may be permitted
to fufpect that experience is not the necef-

x                                         fary

fary appendage of age, or the refult of even an extenfive employment in the difcharge of profeffional duties; nor inexperience invariably the portion of youth, and a more recent initiation into the myfteries of practice. God knows, it is not neceffary to proceed the length of India in queft of inftances to corroborate this fufpicion; for without attention to benefit, by frequent opportunities, and judgment to direct that attention, a perfon of extenfive practice and the moft advanced age will remain for ever an infant in the practical knowledge of his profeffion; but, on the contrary, a practitioner of any age, who poffeffes thofe qualities in a much more confined fphere of practice, will become an experienced phyfician without a wrinkle.

The writer is fenfible, that the freedom with which he has treated thofe refpectable authorities (by no means refpectable when they influence practice in India) may render him obnoxious to the general reproach of prefumption, until the proofs which warrant that freedom fhall be brought forward. As far as this opinion may interfere with the general adoption of the practice recommended in thefe fheets, he will fincerely regret it; in no other refpect can it poffibly affect him.

# DYSENTERY. *

As the fubject of fevers has been treated much more diffufely than was intended, and, probably, than the nature of this addrefs will warrant, the remarks on the fubfequent articles muft be as concife as poffible; the writer will not therefore obtrude on the attention of gentlemen, in any part of thefe remarks, the hypcthetical fuggeftions of others, or the fpeculations which have fometimes occurred to himfelf. Not only the limits of this paper, but the opinion entertained of the ingenuity as well as the leifure of medical gentlemen in the honourable company's fervice, muft preclude fuch difcuffions, and oblige him to confine this article to the expofition of a few practical facts. From thefe, indeed, gentlemen who have been more recently converfant in the fyftems of univerfities and books, are competent to derive plaufible theories with fuperior advantage.

* This article, with fome trifling alterations, was publifhed in Bengal, at the commencement of the year 1789.

It

It is obfervable, that although the climate of Bengal be deemed extremely conducive to the production of fevers and fluxes, yet the higher orders of Europeans experience an extraordinary immunity, in the early ufe of inteftinal evacuations, and immediate recourfe to medical advice, on the flighteft appearances of diforder. It will, no doubt, afford every medical gentleman, on his firft admiffion into the honourable company's fervice, as his emoluments do not depend on the number of his patients, the moft fenfible pleafure to be informed, that, with the advantage of proper treatment, dyfentery, as well as fever, is marked by a particular exemption from danger in thofe regions, and that both thefe diforders are of very trifling duration, unlefs the inattention and ignorance of the practitioner, or the negligence and obftinacy of the patient, prolong their exiftence.

In that climate, every defcription of recent inteftinal flux may be efteemed to arife from fimilar caufes, and to require fimilar means of cure. Notwithftanding the great Doctor Cullen's view of this fubject, and the opinions of many celebrated practitioners, any diftinction between diarrhœa and dyfentery will in general be found unneceffary, if not pernicious. If a fimilarity of caufes and of

H 3 treat-

treatment may be deemed to conftitute difeafes of the fame kind, gentlemen muft receive, from their practice in India, ample teftimony in favour of Sydenham's obfervation,
that the dyfentery is a fever of the inteftines,
and that the obfervation is by no means confined to any one particular kind, but affects
every fpecies of the diforder. The influence
of this opinion fhould confequently pervade
the practice of medical gentlemen in Bengal,
as well as in the honourable company's fhips,
and fhould fuggeft an argument in favour of
the little diverfity of treatment requifite in
fevers and fluxes.

It is remarkable to what a degree of fimplicity the farrago of European prefcription
has been reduced by the general practice of
gentlemen in that country, in all diforders.
In this particular complaint, although the
remedies are fimple, yet the indications of
cure are often erroneous. To very few of
them, however, has venefection, on the bold
prefumption of the exiftence of an inflammatory affection in the inteftines, appeared of
probable advantage, or exempt from danger.
The commencement of thefe diforders is not
unfrequently attended by fallacious fymptoms, indicatory of inflammation; yet a pofitive affurance may be given, that venefection

is

is utterly inapplicable to any ftage of this
diftemper, under any appearance or prefiure
of fymptoms.

Sudorifics, in direct contradiction to their
boafted effects in the Weft Indies, have fometimes proved greatly detrimental, even under
the favourite form of antimonial anodyne
draughts; but towards the termination of
the diforder, if it prove very obftinate, after
the ufe of evacuations from the bowels, they
may be allowed a trial in Doctor Mofely's
method.

The efficacy of ipecacuan, to the exception
of its emetic powers, is too queftionable to
authorife any reliance on its virtues in thefe
complaints. It has alfo, on all occafions,
appeared to poffefs lefs activity in India than
in Europe, on a comparifon of its effects
in practice with the accounts and commen-
dations contained in authors.

The folution of white vitriol, which has
received fuch lavifh encomiums from Doctor
Mofely, will often be found of confiderable
fervice: although it muft feidom be expected
to effect a complete cure, it has generally
operated with efficacy one or both ways, and
in this manner has proved beneficial. As an
aftringent, like all other aftringents, it deferves the moft pointed reprobation; but it

H 4                                        will

will not eafily be perceived to act in this
manner; for the agitation of naufea, by this or
any other means, is generally calculated to
promote the natural evacuations from the
bowels.

It is fcarcely neceffary to apprife gentlemen, that alexipharmacs, as expreffive of thofe
multifarious compounds of difpenfaries, are
a word of little meaning, and medicines of
little ufe; that they cannot be fuppofed to
retain any degree of prefervation in warm
climates; that in cafes which may be thought
to require their fancied powers, fubftitutes of
a lefs fallible operation are greatly preferable,
from their fimplicity; and that, in the difeafe
under confideration, they will generally aggravate the fymptoms.

Medicines, of effects fomewhat fimilar,
combined with ipecacuan, have proved fa-
fhionable in Europe. Perhaps no part of
European routine deferves, in moft inftances,
fuch decided rejection. The high character
of thefe medicines renders the moft particular cautions on this fubject neceffary. Their
exhibition may be pronounced fatally trifling
in the firft ftage, and of very doubtful agency
during any other period of the complaint.

The general prejudices in favour of bark,
in a very numerous body of difeafes, fhould
make

make the writer exprefs himfelf with peculiar hefitation on this fubject. To fpeak,
however, with a degree of doubt which he
does not entertain, and a diffidence which he
cannot feel, may be a plaufible refinement
of fcientific hypocrify, but would prove a
very imperfect declaration of his real fentiments; yet a confidence, which naturally refults from the teftimony of a moft fuccefsful
practice, might appear to border on empiri-
cifm. Moft authors have recommended bark
in fome ftages of the difeafe, and many have
deemed its ufe indifpenfible, if not infallible,
in certain imaginary tendencies to putrefcency.
But we need not fcruple to believe that bark
is inadmiffible under any circumftances, or
during any period of the actual exiftence of
this diftemper. Technical expreffions of un-
determinate, or of no meaning, have at all
times proved the refource of ignorant men,
and may have occafionally mifled the moft
enlightened. When a vague fufpicion of
*relaxation, atony, putridity*, has glanced on the
practitioner's imagination, a ready expedient
has prefented the numerous lifts of *corroborants, tonics, antifceptics*. In this ftrange nomenclature, bark has always obtained the firft
place, and has confequently been exhibited by
fome practitioners, if not the generality, in

§                                                    India,

India, with a liberality not inferior to the utmoft fcope of European favour, though not
with equal felicity, if we may truft our reading
on the fubject of its fuccefs in Europe. Nature,
with her ufual fagacity, has on many occafions
pertinacioufly rejected this favourite drug,
and has been foothed into a temporary acqui-
efcence by the addition of opium.

The retention of other medicines is not the
moft important purpofe for which opium has
been prefcribed. The fufpenfion of pain
muft ever prove an object of the firft confideration and confequence to the practitioner
as well as to his patient, when it can be
accomplifhed without an ultimate aggravation
of all the fymptoms. The operation of
opium, although it may often prove a momentary alleviation, will not be found in the
event to poffefs this immunity; practitioners fhould confequently refrain, as much as
poffible, from its exhibition in any form.

A fimilar reftriction is equally neceffary in
the article of wine. Although the nature of
this diforder exhibits itfelf frequently under
deceitful appearances of great weaknefs and
depreffion of fpirits, yet it will foon be perceived, that unlefs the ftomach reject this
remedy early, it will excite or increafe a con-
comitant fever, and that its ufe is utterly
                                inadmiffible

inadmiffible in the treatment. During a course of practice in the military hofpitals for European foldiers in Bengal, there are frequent occafions to obferve, that patients of every denomination crave an allowance of wine with the utmoft earneftnefs; that the mere name of wine, or liquor of any kind, is fo cherifhed by them, that they will drink it with avidity, under a certainty of its immediate rejection; that the moft fober and prudent will conceal and referve their allowances, until an opportunity occur, to difpofe of the accumulated treafure to their own pecuniary advantage, and the great detriment of the purchafer, who is generally a lefs prudent patient; and that fcarcely one man in one hundred poffeffes the refolution or the honefty to acknowledge that it has proved in any degree prejudicial to his diforder.

The exhibition of thefe fatally vigorous medicines has been the offspring of a jargon (excufe the term) which has long filled the pages of authors. *Relaxation* has been uniformly pronounced the affociate of moft diforders in warm climates, and methods of treatment, adapted to counteract this, particularly recommended, with what degree of fagacity or knowledge the unprejudiced perufal of the *Select Evidences*, &c. which are

now

now in the prefs, may determine. In the private practice of Bengal, as wine is perceived frequently to grow four and return from the ftomach, or to occafion cardialgia and other difagreeable fenfations, fpirits diluted with various proportions of water are often fubftituted in the place of wine. This mifchievous beverage is alfo dictated by a fancied relaxation. The writer may here venture to advert to the very pernicious error of an allowance of fpirits which formerly prevailed in hofpitals. Although the practice has been, with the greateft propriety, abolifhed in Bengal by authority, yet it may poffibly ftill exift in the hofpitals of the honourable company's other eftablifhments, and may be thought to deferve the notice of their governments. That the exiftence of fuch an evil fhould continue to have the fanction of hofpital practice in Europe, or that it fhould be permitted to form a regular part of the diet of European foldiers, as well as failors, in India, is a fubject of aftonifhment, and of the moft ferious condemnation. But this is not the place to exhibit the inexpediency and impropriety of that allowance in its true colours.

The method of treatment, which has proved moft fuccefsful during a courfe of

practice

practice in Bengal, and on board the Houghton, may be comprifed in a very few words.
Experience, fatal in many inftances at the
commencement of the writer's career, at laft
compelled him to infift on inteftinal evacuations with the greateft freedom, and to confine his practice, ultimately, to a repetition of
purgatives and emetics. Solutions of falts,
however, unlefs at the very beginning, to
evacuate, as in fever, the groffer contents of
the inteftines, or, in very flight and recent
cafes, will feldom be found adequate to the
removal of the diforder; and a frequent
reiteration may be attended with great detriment. Caftor oil is certainly a medicine,
in this difeafe as well as in fever, of more
efficacy in the difcharge of bilious, glareous,
and other vitiated matters, and occafions lefs
difturbance and debility. But no medicine
whatever, perhaps, is in any degree compara-
ble to calomel in the copious refults of an
operation equally mild and certain. The
griping occafioned by greatly vitiated and
acrid fecretions in the bowels, and perhaps
by the action of the inteftines, to free themfelves from this load, has commonly been attributed, poffibly from fuperficial obfervation
only, to medicines which poffefs the power to
difengage

difengage and difcharge them. The exhibition of calomel fhould always be followed by other purgatives in a liquid form. The fenfe of weaknefs, which immediately fucceeds thefe evacuations, and poffibly all inteftinal evacuations, even the moft natural, fhould not deter the practitioner from a bold repetition of the fame means; for the patient will be ultimately found to acquire ftrength. The occafional interpofition of emetics will, in all obftinate cafes, prove abfolutely neceffary.

On the fubject of chronic fluxes, which are always, like flow fevers, the confequence of neglect, or mifmanagement, little may be faid at prefent. It may be declared very pofitively, once more,. that they are, in almoft every inftance, occafioned by neglect, or, what is ftill more pernicious, the unfkilful treatment of preceding diforders; even thefe, unlefs in the very laft ftage, admit of moft relief from inteftinal evacuations. In fome rare inftances, opium has been thought to prove ferviceable, in combination with a large proportion of calomel; by which means the qualities of each, that might be fuppofed moft inimical to tender, yet torpid, bowels, would appear to be reciprocally corrected.

An

An alterative courſe of mercury (we need not vouch for the accuracy of the term) has ſucceeded very often in theſe fluxes, eſpecially in caſes which have been ſuppoſed to proceed from obſtructions in the liver, or other bowels, and in which the ſecretions of thoſe parts are always greatly deranged. The ſucceſs of this courſe is greatly promoted by an occaſional exhibition of emetics and purgatives.

Proofs of the ſucceſs of theſe methods cannot be added in the narrow compaſs of this paper; yet ſuch is the conviction the writer entertains of the deleterious agency of wine, bark, opium, and aſtringents, in the fluxes, as well as in the fevers of thoſe climates, that, however unuſual or abſurd an appeal to the feelings of ſurgeons may ſeem, he will venture to intreat them, in the moſt earneſt manner, as they regard the lives of thoſe entruſted to their care, and their future reputation, in that country at leaſt, to refrain altogether from the exhibition of thoſe medicines, until they ſhall have peruſed the public proofs of the ſucceſs, or ſhall have experienced the inefficacy of the means, recommended in the foregoing pages.

L I V E R.

# LIVER.

THE fuccefsful treatment of this diforder, or rather variety of diforders, has long
been underftood in the honourable company's
fettlements in the Eaft Indies, and perhaps
there only. This fortunate circumftance
might feem to render a minute difcuffion
of the fubject unneceffary, efpecially in an
addrefs of this nature. However, though the
great outline of practice in the liver is equally
well known to the medical gentlemen in
India, and in the honourable company's fhips;
though it is not a fecret to any European of
any profeffion in thofe climates, that mercury, in almoft every form, is a pretty certain
cure in complaints of this vifcus; and though,
in many inftances, the fymptoms defcribed in
books are fo evident as to render a mifappre-
henfion of the nature or the feat of the difeafe
nearly impoffible, yet fome errors in the con-
duct of the cure, fome mifmanagement in the
exhibition of this admirable remedy, prevail
fo generally, as frequently to retard the completion of the former, or to render the operation of the latter utterly abortive. While it
is acknowledged, with great fatisfaction, that
the

the treatment of this diforder, during its early ftages, and in its moft obvious forms, proves almoft always fuccefsful in that country, it may be afferted, that the liver gives birth to an infinite variety of fymptoms, the fource of which is entirely unfufpected by the generality of medical men; that the difeafes formed by various combinations of thofe fymptoms are much more numerous than the more obvious affections of that part; that, as the caufes are miftaken, the difeafes themfelves are mal-treated; and that, confequently, no diforder whatever of thofe latitudes would feem to demand more attention, and more minute inveftigation, than thofe which, by the nomenclature of the eaft, may be collectively denominated *The Liver.*

It is not poffible, in the fcope of this paper, to do any manner of juftice to fuch an extenfive fubject. It well deferves a feparate and enlarged treatife. Very curfory and general remarks are the utmoft that can poffibly find admiffion here. The writer, however, will attempt to make thefe include what is moft effential and ufeful, in the materials which are at hand. The fubject may be arranged under two general heads; the fymptoms which afford a fufpicion of, or decidedly characterife, an affection of thofe parts, and

I                                        the

the remedy, which ftands foremoft and almoft alone in the treatment, or mercury. The former is by far the moft important, at leaft generally known. Although much remains to be difcuffed in the latter, yet, as it generally fucceeds under any circumftances of exhibition, it may be allowed a fmaller fhare of attention here.

SYMPTOMS.—The annals of phyfic, or the medical fafhions of various periods, afford fufficient proof, that particular terms, fometimes expreffive of fome meaning, but much oftener of none, have each been in univerfal vogue for a time, and, during their reign, have been fuppofed to give character to almoft every diftemper that occurred. In this manner, a race of inflammatory, bilious, putrid, nervous monarchs fucceeded one another. To render this ftrange fucceffion confiftent with the writings and reputation of phyficians, who lived under the different reigns, and to reconcile the apparent contradictions of them all, later writers have fought for the caufe of this diverfity, in the variety of diet, modes of exercife, and other habits of life, which were deemed to characterife each period. Thus the maids of honour in Queen Elizabeth's court, from their breakfafts of

beef-

beef-fteaks and porter, were naturally fuppofed more obnoxious to inflammatory and putrid diforders, while thofe of Queen Char-
lotte's, from their exceffive ufe of tea, prove
equally liable to nervous affections. It is at
prefent the reign of nervous; and this term,
notwithftanding Fontana's inimical experi-
ments, is fo unintelligible, fo perfectly well
calculated to puzzle and to pleafe the vulgar
of all degrees, and to impofe on the too inqui-
fitive patient, that its dominion is not likely
to prove of fhort duration. Far be it from
us to difturb its peaceable reign over countries in which we have no experience, and
patients in whom we have no concern. But
as this unmeaning term has been obferved
fatally influencing practice in Bengal, and
exhibited as characteriftic of fymptoms, which
certainly originated from a fource connected
with liver, we ought to contribute our beft
endeavours to exclude it entirely, or with very
little refervation, from all fhare in the treatment of diforders in that country.

It would fcarcely be exaggeration to declare, that not only the tribes of nervous
fymptoms, but thofe of almoft every other
character, appear, in fome manner or other,
either as caufe or effect, to be unaccountably
connected with liver, in the moft enlarged

I 2 acccptaacceptation of that word, in India. This appellation is, therefore, by no means confined within the limits which the name of the viscus would feem ftrictly to require. It was formerly, in Europe, pretty much the practice to fuppofe every difeafe, curable by mercury, connected with a venereal taint. In this refpect, liver has fupplanted the lues venerea in India, and with great propriety, for every purpofe of practice. It may be plaufibly alledged, that while the European extenfion of a term to all diforders is condemned, we incur a fimilar charge by the very general application of liver to the complaints of other latitudes. It is certainly no very uncontrovertible argument to fay, that becaufe mercury has been found of almoft certain efficacy in affected liver, and has proved equally fuccefsful in a great variety of other difeafes, that thefe difeafes muft neceffarily have been the liver. On all occafions of fuccefs, precarious conjecture alone muft decide the nature of the diftemper; and in this view, the beft practitioners will neceffarily have the feweft opportunities of forming a juft judgment. But as the very beft are not always infallible in their practice, frequent occafions will occur of afcertaining the real nature and fource of the previous fymptoms,

almoft

almoft beyond the poffibility of deception, by
luminous diffections. In this confifts the advantage over the European application of general terms. Diffections, it is apprehended,
afford very fallacious information, notwithftanding all that has been written on the
fubject, refpecting the exiftence of nervous
caufes exciting the previous difeafes. They
may not be altogether decifive, even in bilious, putrid, or inflammatory cafes. But
when diffection exhibits the vifcus in queftion, deranged in its form, fubftance, fize, or
even offices, beyond all expectation, and almoft beyond conception, when the other
bowels, or contained parts of the body, are
either entirely exempt from morbid appearances, or feem affected in a trifling degree,
will it be poffible to refufe our affent to the
conclufion, that an affection of the liver was
the caufe, the confequence, or in fome manner connected with the fymptoms, which
feemed to put a period to the patient's exiftence? Such, therefore, is the nature of the
evidence on which the general connection of
liver with all complaints in thofe countries
is indifputably founded. Diffections are not
fufficiently numerous, perhaps, in any country; certainly not in that. The heat of the
climate, and confequent rapidity of the putre-

factive

factive procefs in bodies, prohibit the frequency of diffections. Favourable opportunities, however, have not often been neglected; and thefe, as far as the writer's affiduous enquiries have extended, have invariably afforded teftimony of the exiftence of
morbid appearances in all livers, which have
undergone examination, whatever the preceding diforder may have been fuppofed. His
individual obfervation has not extended to
very many inftances of ocular evidence; but
no proper occafion of confulting the experience of others has at any time been omitted,
and all confirm the reality of this connection.
Thus it appears, there can be no great impropriety in the application of this term to
a great variety of diforders, as the vifcus is
found fo generally affected. But for the ufeful purpofes of practice, the term, as influencing the method of treatment, fhould include a larger clafs of diforders of very doubtful origin, or of no obvious affinity whatever
to that vifcus; and provided the method of
cure be fuccefsful, it will not, perhaps, be
deemed a matter of very great confequence
to afcertain the exact feat and nature of the
late diftemper.

On all occafions, when the fecretion of
bile is morbidly increafed, diminifhed, or irre-

3

gular,

gular, it is clear, that the vifcus has not its
natural action, and does not perform its proper offices. Thefe inftances alone would
affect almoft every diforder in that country,
but efpecially fever and dyfentery, moft particularly in their chronic forms. The fecretions, however, of the other bowels, are very
generally deranged in a fimilar manner, and
often at the fame time as the former; and it
would be as difficult as unneceffary to give
each its particular fhare of fymptoms in the
diforder; but the liver fhould be fufpected
under every derangement of the fecretions of
any of the bowels. Thefe appear to have a
concern in all chronic complaints in that
country; in old pains, general or local, that
are not decidedly venereal; in moft, or in all,
cuticular affections; and in all obftinate complaints of the head, cheft, and abdomen.

The writer will here enumerate, without
attention to order, fuch fymptoms, affecting
each of thofe parts, as occur to his recollection
at prefent. Head-ach, in every kind of form,
is by no means an uncommon effect. Thofe
which are called nervous will be always
found connected with the fource in queftion.
Periodical head-achs are frequent; thofe
which recur with obftinacy on rifing in the
morning are very fufpicious. Obftinate or

periodical

periodical pains in either or both ears are by
no means of rare occurrence. Frequent and
fevere thirft, if this be referable to the head;
a bitter, or other unpleafant taftes in the
mouth; fœtid breath, efpecially in the morning; various difcolouration of the countenance; fometimes yellow, fometimes extremely free from all yellow tinge, pale, but
not clear, often of a deadly leaden colour,
fometimes of a copper tinge, on fome occafions a kind of mixture of all thefe in various
proportions; frequent flufhings of the whole
face, or of particular parts only, efpecially of
a circumfcribed fpot in one or both cheeks;
partial fweats of the head, fometimes of the
forehead only; frequent giddinefs, or confufion, on ftooping particularly; fenfe of weight
in the head, fometimes approaching to a degree of pain, efpecially on moving fuddenly,
or on rifing on the toes, and falling on the
heels again with a jerk; frequent drowfinefs;
mouth generally foul and covered with vifcid
fecretions; the fauces particularly foul, and
often fœtid; the gums and other parts of the
mouth liable to frequent ulcerations, which
are extremely painful when touched, but not
otherwife; the gums fometimes fufficiently
firm, at other times apparently deficient in
firmnefs and colour; the teeth alfo lefs white,
and

and much more difficult to keep clean, fubject to frequent aching, and to become cari-
ous; the falivary fecretions irregular; the
tongue very feldom perfectly clear, but of all
degrees and complexions of foulnefs, though
generally moift, fometimes, however, very
dry after fleeping, and fubject to ulcerations like the other parts of the mouth; a
remarkable degree of inaptitude to continued
motion in the lower jaw, particularly evinced
in chewing; the eyes fometimes yellow, generally muddy, and on fome occafions perfectly free from all tinge, clear, but not
vividly clear as in health, and fometimes of
a flight bluifh caft; vifion not quite fo perfect as in health, efpecially after looking for
any time at one object; in fome inftances, a
general greafinefs of the face, and a glaffy appearance of the eyes; ftuffing of the nofe,
and other catarrhal fymptoms, fometimes a
difcharge of blood, without any apparent
caufe. All the fymptoms, which I have juft
noticed, will be found to attend affections of
the liver in various proportions and modes of
combination on different occafions; but they
are in general more decidedly characteriftic
of that ftate of the fecretions of the bowels,
which, in Bengal, and probably all over India,
is termed *The Bile*; an appellation which includes

cludes the morbid increase, diminution, or alteration of any of the secretions of those parts, consequently the state of the digestive powers, producing unpleasant sensations, more or less severe, according to the degree of deviation from their natural state. *The Bile,* then, although frequently unconnected with any permanent affection of the liver, or any other part, and consequently curable by less active means, such as laxatives, diet, and exercise, is almost always a pretty certain attendant on fixed derangement of the liver, even when this does not appear to form the smallest quantity of its peculiar secretion.

As *The Bile* will be noticed occasionally under the general title of *The Liver,* we may dismiss it for the present, and proceed to the enumeration of symptoms affecting the thorax.

Perhaps there is no variety of derangement in the functions of the lungs, but what may, and does occasionally, originate from the liver; but this is not the proper place to take notice of acute cases. All denominations of asthma, dry, humid, nervous, have often proved symptomatic of the liver, both in the strict and extended application of the term. Every alteration of the offices of the lungs, from the natural state of respiration, from a slight sense of

of impediment to a condition not far removed
from fuffocation, increafed, fudden, quick,
difficult, laborious refpiration, are all generally charaƈteriſtic of the liver. To thefe are
added a greater difficulty of breathing after any
effort; a fenfe of fuffocation occafionally; a
perception of heavy weight preffing on the
lower parts of the lungs; a cough in all its
varieties, generally dry, for a length of time
at leaſt, frequently conftant, often recurring
only at intervals, fometimes very flight, fometimes the principal and only fymptom, when
moiſt accompanied by excretions of every
colour and confiſtence; pain in one or both
fides, or in other parts of the cheſt, fometimes
permanent and apparently pleuritic, at other
times intermittent or periodical, often of a
peripneumonic charaƈter; difficulty of laying
on one or both fides, and fometimes of a recumbent poſture altogether, though not often
in chronic cafes, chiefly, however, in the laſt
ſtages; a fenfe of ſtriƈture about the pharinx,
or of weaknefs there; fometimes a flight degree of difficulty in fwallowing; a degree of
hoarfenefs, and various changes in the voice;
a palpitation of the heart is by no means uncommon. When any number of the preceding fymptoms prove obſtinate, it will
always be prudent to have the liver in view.

Thofe,

Thofe, which would feem to originate from
the abdomen, are much more decifive as well
as more numerous; they may be mentioned
under general heads.

Dyfentery ftands foremoft, in all its varieties, from a flight diarrhœa to its worft
forms of gripings, ftrainings, and difcharges
of blood and other matters. The appearances of the ftools are as follow: flimy,
glareous, green, yellow, bloody, black, clear
and watery, brownifh yellow, fometimes thinner, fometimes thick, frothy, windy, deep
yellow, fœtid and watery, yellow flime depo-
fited in thinner ftools, thick mucus tinged
with blood, dark brown fœtid mucus, a diffufion of blood and bile, a little white matter
with blood, often pretty natural, fometimes
thick and brown, fometimes great coftivenefs,
fometimes frequent alvine dejections, fcarce
and fmall without fcent and difcolouration,
white like cream and liquid, fometimes like
the wafhings of raw meat, fometimes mixed
with black blood or black bile, grumous
blood mixed with various matters, particularly philaments, bloody ftools without pain
or fibrous matters occurring, after long intervals, in large quantities, fometimes purulent,
colliquative diarrhœa, attacks of cholera, large
and irregular difcharges of bile, and a total
or nearly total deficiency in the intervals.
The

The ftomach is alfo affected in a great variety of ways. The difcharges by vomit,
either fpontaneous, or excited by medicine,
are equally various; fometimes of mere
phlegm, four- and faltifh water, dark green
nearly black with pieces of a blue fubftance
like tenacious mucus, acid, bitter, fweetifh,
taftelefs; fometimes pure bile of various
colours, from a bright yellow to black, frequent or inceffant naufea, frequent or inceffant vomiting, wind accompanying other
matters or alone, pains in a great many
forms attack the ftomach; fometimes a
gnawing at the orifice, often a fixed pain
at the pit of the ftomach, uneafinefs and
forenefs about the lower part of the ftomach
and fternum, unpleafant heat in the ftomach
on fwallowing wine, fpices, and other things;
fometimes entirely free from all pain or uneafy fenfations, fulnefs at the pit of the ftomach, with or without pain, fhooting pain
through the regions of the ftomach and
fpleen, or from the epigaftrium to the back,
pain from wind occafionally, cardialgia, and
numerous other pains, apparently of the ftomach itfelf. It is very remarkable, that in
the moft decided affections of the liver, there
is often neither pain nor uncommon fenfation of any kind in the part itfelf. In very

acute

acute cafes indeed, both pain and fwelling
commonly make their appearance to a great
degree. Pain, however, in this part, is feldom very fevere; but it is of very various
characters when it does occur. It is moft
generally obtufe, and very obfcure, fometimes
recurring after brifk motion, or after meals
only, fometimes permanent. It attacks every
part of the liver, confequently the pain is
occafionally pretty high up under the ribs,
at the edge of the falfe ribs on either fide,
extending fometimes from the pit of the ftomach in the courfe of the falfe ribs of the
right fide to about the region of the kidney.
A fenfe of fome fulnefs and an uneafy fenfation, conftant or occafional, is much more
common than pofitive pain; but it is ftill
more ufual to have no extraordinary fenfation
whatever in the part itfelf to point out the
feat of the diforder. The pain is not uncommonly felt only on the left fide; and
indeed the liver, in a morbid ftate, extends
far on that fide fometimes. A general forenefs infide is not unufual, without much pain;
in many inftances, no pain whatever is felt,
even on preffure, in any part of the abdomen.
It attacks the neighbourhood of the cartilago
enfiformis, in the form of an infupportable
ftricture fometimes; and fometimes fevere
pains

'pains are felt, as if the bowels were wrung or pulled down with a hand. They are fometimes extended externally over the abdomen, and fometimes internally, or round the umbilicus; excruciating, and fhooting in every poffible direction, from the lower belly up to the right fhoulder, the whole abdomen being tenfe, hard, and extremely painful to the touch, as well as much inflated; the belly at other times foft, though fwelled, with pain on preffure, and hardnefs round the navel. Weight, without pain, occurs in various parts of the belly, but moft frequently in the region of the liver, often accompanied by different degrees of oppreffion of the præcordia, efpecially after food. The belly in many inftances is confiderably funk, and often apparently of a natural form, without a fingle circumftance that indicates diforder there. The region of the bladder is often much affected, and the urine exhibits every variety of condition; it is fometimes exactly as in health, or of a beer colour, or pale, clear, in fmall quantities, white, muddy, experiencing. fudden alterations, like Madeira wine, difficult, hot with ftraining, fuppreffed, like muddy porter, greatly tinged with bile, or totally free, and all the intermediate ftages thin and watery, fometimes fœtid, fometimes

apparently

apparently purulent, much sediment or none, perfectly clear or much clouded, depofitions of various kinds, fometimes bloody. The fpleen is often very obftinately enlarged, but feldom with any confiderable fenfe of pain. Prolapfus ani, but particularly the piles in every ftate of them are very common; and violent hæmorrhages, fometimes unconnected with the piles. Dropfy, in all its forms, is very frequent. Flatulencies both ways occur often, and hiccough fometimes, as well as violent eructations. Thefe are the principal fymptoms that affect the three cavities; there are many others, no doubt, which do not at prefent occur to his recollection. There is fcarcely any local complaint whatever, but is occafionally imitated by the liver; and all old local complaints are very apt to be affected by this diforder. The gout is fometimes connected with it; the jaundice, in flighter degrees, very frequently, although a violent jaundice is not, probably, a very common diforder in India. Temporary fuffufions often occur, but the permanent jaundice is rare. Chronic rheumatifm would appear to have a confiderable connection with liver complaints. Local pains are, indeed, very common. The pain, however, in the fhoulders, in one or both, though it often occurs,

parti-

particularly in acute cafes, is much more frequently abfent. But he would never come
to a conclufion, were he to enumerate every
fymptom that has been found to proceed
from, or to be connected with the liver. He
muft not, however, difmifs the fubject of the
fymptoms, without a few obfervations on
fevers as connected with this vifcus.

There is no one form of fever in Bengal,
but what has occafionally been found connected with the liver; the moft violent and
continued kind would feem to accompany the
acute affections of the liver, and the flow and
intermittent fevers, chiefly the chronic difeafes of that part. The writer has obferved
before, that liver very often proceeds from
the mal-treatment of preceding fevers; and,
no doubt, it is often the original difeafe, and
the caufe of fevers of great obftinacy. It
will be highly proper to pay great attention
to the ftate of the liver in all fevers of warm
climates. But we may conclude thefe curfory remarks on the fubject, with fome account of its connection with the yellow fever,
or black vomit, of the Weft Indies. During
the rainy feafon of the year 1789, feveral cafes
of the yellow fever, as defcribed by Weft
India practitioners, and by Doctor Lind, occurred in the general hofpital at Calcutta.

K                    The

The patients, previoufly to the attack, had inhabited the low, damp, arched ways of the fort. Amongft other fymptoms, the fkin was particularly fuffufed with yellow, and lumps of black bile, apparently, were difcharged by vomit. The ftools were black and putrid. On the firft appearance of this diforder in the hofpital, a medical gentleman intimated, that during one of his voyages to the Eaft Indies, he had thirty patients at one time afflicted with this difeafe. He opened the firft who died, and found the liver fo particularly affected, that he immediately exhibited mercury, in various forms, to the others; all of whom recovered. On opening the firft patient who died in the hofpital at Calcutta, the liver was found aftonifhingly enlarged, afcending as high as the fecond or third true rib, and replete with matter. The lungs, in the right cavity of the thorax, were compreffed by the enlargement of the liver, into a very fmall and hard mafs; the cavity itfelf, confequently, almoft obliterated; yet it is obfervable, that this patient had not been affected with any difficulty of breathing. The fpleen was alfo greatly enlarged, and full of matter. The medical gentleman, who had treated the cafes on board the fhip, obferved, that in the former the fpleen was alfo greatly

4                                              enlarged,

enlarged, and, on diffection, a large quantity
of blood flowed from it. On this difcovery,
the other patients in the hofpital at Calcutta
were immediately put under a courfe of mercury internally and externally, with James's
powders. The mercurial inunction, without
intermiffion, was applied to every part of the
fkin where abforption could be expected to
take place, and continued until the mouth
began to be affected, when every fymptom of
the diforder vanifhed; conftant purgatives or
laxatives were ufed during the whole courfe
of the difeafe, and emetics were occafionally
exhibited, with great relief to the patient;
nor did they, in any inftance, induce that exceffive irritability of the ftomach attributed
to them by practitioners in the Weft Indies.
This exemption may reafonably be fuppofed
the refult of previous purgatives, which had
evacuated the contents of the duodenum and
neighbouring parts, and had given the putrid
bile and other acrid matters their natural tendency downwards. The violence of the fever
was foon fubdued by the purgatives; but the
difeafe did not entirely difappear until the
mercury took effect. Wine, exhibited after
the fever had been reduced, was thought to
relieve very effectually the oppreffion and
anxiety about the præcordia, to enable the

patient

patient to fupport the conftant exhibition of purgatives, and to prove greatly reftorative in other refpects. The good effects of wine appeared fuch to all the gentlemen who attended thofe cafes. The bark was entirely neglected, unlefs as a ftrengthener after every fymptom of the difeafe had vanifhed. It appeared highly proper to introduce this circumftance here at more length than the limits of the paper would feem to admit, as there has fcarcely at any period occurred a more important fact in the records of phyfic. How many lives may be faved in the Weft Indies by the knowledge of this fact may be conjectured from the opinion of the great fatality of the yellow fever amongft the foldiers and failors in that part of the world.

Women and children are very liable, in thofe countries, to obftinate derangements of the fecretions of the bowels, which give rife to a variety of peculiar complaints, fuch as marafmus, atropia, &c. all curable by the fame means. This, however, is not the place to enter on the fubject of their diforders; we may therefore only obferve, that it is incredible what a quantity of mercury may be given to both women and children in thofe climates, and in how great a variety of their complaints, not only with perfect innocency, but with great benefit.

It

It will be proper to exhibit here fome few of the appearances on differtion. In thefe there is no regularity or conftancy. The fame ftate of the liver does not, by any means, feem to occafion, invariably, the fame fet of fymptoms; and, confequently, any particular combination of fymptoms will feldom lead to the knowledge of the particular condition of that vifcus. In reality, this accurate knowledge would appear quite unneceffary in the general treatment; for every fet of fymptoms arifing from this fource will be found to yield to fimilar means. Nothing can be more various than the fize of difeafed livers. They are often fwelled in an enormous manner, occupying the left as well as the whole of the right fide, and extending upwards in fuch a way, as almoft to annihilate the lungs and the cavity of the thorax; and, on thefe occafions, it is very extraordinary, the breathing will fometimes remain apparently natural, while on others, when the fize of the liver is aftonifhingly diminifhed, which is often the cafe, the breathing is uncommonly affected, even when the differtion can difcover no traces of difeafe in the lungs themfelves. In both the morbid enlargement and diminution of its fize, the texture will fometimes appear

K 3                    perfectly

perfectly natural. The variety of morbid appearances, which take place in the texture of the liver, is very great. Sometimes one large abfcefs, fometimes a number of fmaller abfceffes occur without any other marks of difeafe. Thefe are occafionally combined with others. Livid fpots, a total confufion of texture, coagula of blood, blood in a liquid and perhaps putrid ftate, white firm fub-ftances like fchirrous glands, are amongft the number of appearances. The colour is very various, often perfectly natural, and often directly the reverfe. No particular number of thefe are conftantly found com-bined. The combinations occur in every poffible variety; and thefe, which have been enumerated, are, indeed, a very fmall part of that variety. The liver alone is often af-fected, when no other part of the body exhi-bits figns of difeafe; although various local pains, during the courfe of the complaint, may have led to a fufpicion of particular parts be-ing affected. The lungs are often in this pre-dicament, as well as the ftomach. The former are fometimes much reduced in fize, or of the natural fize, but with their texture more or lefs deranged on one or both fides; fome-times they are marked with livid fpots only, fometimes with bloody extravafations; on

other

other occafions, abfceffes take place of various
extent, or of fchirrofities of a whole lobe, or
of a number of little fpots, or a total deviation from the natural colour and texture.
The various appearances of the lungs, on diffection, which in Europe are deemed characteriftic of different ftages of confumption,
will be found to attend liver complaints, and,
in general, would feem to be curable by the
fame means, as far as a judgment can be
formed from fuccefsful cafes. Although the
functions of the ftomach are generally deranged in every variety of way, during the
exiftence of the diforder, yet it does not
often exhibit any confiderable morbid appearances on diffection. It is fometimes much
diftended, and fometimes reduced extremely
in fize; its coats are occafionally thickened,
and at other times fhew fpots of extravafated
blood; the colour alfo is various, and it will
not always be found in its natural fituation.
The fpleen is often not at all affected, but
occafionally its texture, colour, fituation, and
fize, are as confiderably and as varioufly deranged as the liver itfelf. Sometimes it has
been found a mere collection of putrid blood,
and other matters. It is fometimes enlarged
to a great degree, without any other morbid
appearance; and has occafionally abfceffes,

K 4 fchirrho-

fchirrofities, and other marks, which at any
time occur in the liver. The gall bladder,
as well as the ducts, undergo a great many
changes. The former is often turgid, with
vifcid black bile, fometimes greatly enlarged,
fometimes empty, and much fmaller than
natural. The ducts are enlarged or dimi-
niſhed in fize, their coats thickened, or their
channels entirely obliterated. The inteftines
do not often exhibit very pofitive figns of
difeafe; extravafations of blood, however, occur in their coats; they are fometimes much
collapfed, and fometimes much diftended;
the colour is often not perfectly natural, and
the infide is fometimes lined with tenacious
mucus, or vifcid bile of various colours. The
mefenteric glands are fometimes greatly difeafed in chronic cafes. The omentum is
either entirely in its natural ftate, or much
reduced in fubftance, or entirely obliterated.
The heart is generally found, though extravafations of water, generally bloody water,
take place in the pericardium, the cavity of the
thorax, or the abdomen; extravafations alfo
of blood and matter are found in the two latter. Although, during the courfe of the
preceding diforder, both the kidneys and the
bladder may have feemed confiderably affected, yet it is not often that any figns of
difeafe

difeafe are difcoverable in them. From this
fhort and general fketch of the appearances,
on diffection, we may proceed to the method
of treatment, which will be made as concife
as poffible, after noticing that adhefions of
various parts are of frequent occurrence.

TREATMENT.—On all occafions of apparent liver and bile, where there is no fixed
derangement, regimen will be found of fufficient efficacy. In great numbers of cafes,
accumulations of putrid bile, and other vitiated matters, will at firft give the ftrongeft
fufpicions of a permanent affection of the
liver, or of fome of the glands of the abdomen; but as the treatment of thefe and of
the liver is exactly the fame at firft, the miftake cannot, in general, be attended with any
bad confequences. Thefe vitiated accumulations in the vicinity of the ftomach, fometimes, as was before obferved, excite a violent fever, with every fymptom of an acute
affection of the liver. Blood-letting, on fuch
occafions, is quite unneceffary, and may be
very pernicious; fo that it fhould never be
ufed, until the real nature of the complaint
fhall have been afcertained by purgatives.
In this, therefore, and in all cafes of the bile,
or incipient liver, the firft ftep is to evacuate
the

the bowels completely, firſt with gentler
ſaline medicines, and then with the more active purgatives, until every appearance of
ſordes vaniſh from the ſtools. Theſe will
in general be found to carry off all, or nearly
all, the ſymptoms; and then a mild vegetable
diet, with water for conſtant drink, accompanied by the moſt regular courſe of exerciſe,
principally on horſeback, will commonly re-
ſtore the patient to his uſual health, or pro-
tect him, with the aſſiſtance of occaſional
laxatives, from any very violent attacks of the
diſeaſe in future. After the firſt operation of
the purgatives, the occaſional intervention of
emetics may be often neceſſary. If the
ſymptoms continue obſtinate after theſe evacuations, the caſe muſt be conſidered as an
*affection of the liver*, and treated accordingly
with mercurials. Whether it be merely a
collection of bilious ſordes, or a real diſeaſe of
the liver, the purgatives and emetics will be
infallibly found to diminiſh the violence of
the complaint, and to carry off many of the
ſymptoms. If the more violent local ſymptoms do not abate conſiderably after the firſt
copious evacuations, veneſection is then ad-
miſſible; a repetition will ſcarely ever be
found neceſſary, where the inteſtinal evacuations have proved abundant. Bliſters are
gener-

generally applied over the feat of the pain,
where this is at all acute *. Practitioners
may be very eafily deceived with refpect to
the abundance of the difcharges by ftool.
They may be affured of the neceffity of further purgatives, as long as fordes of any kind
appear in the ftools, or as long as they are
confiderably tinged with bile of any colour.
After the operation of a medicine, the laft
ftools may be pretty free from both; but on
thefe occafions, other fecretions, other accumulations, will immediately form, and the
next purgative will often produce the moft
copious difcharges of bilious matters, vitiated
mucus, &c. which do not always appear to
be formed of fecretions which are perfectly
frefh.   On the leaft recurrence of the fymptoms, fuch as weight, pain, fulnefs, or oppreffion about the liver and ftomach, the
purgatives fhould be repeated, not only in
mere accumulations, but during the whole
courfe of treatment for a confirmed affection
of the liver.   When, from the obftinacy of
the fymptoms, there is reafon to fuppofe a
morbid ftate of the liver itfelf, or of the
fpleen, mefenteric glands, or other glandular

* It was thought neceffary to repeat fome of thefe obfervations in the Remarks on the *Inftructive Failures.*

parts

parts of the bowels, or that the fecretions of any of thefe are permanently deranged in any manner whatever, recourfe fhould be immediately had to mercurials.

In the flower and chronic forms of liver, a courfe of mercury in any way anfwers, perhaps, equally well. The common mercurial pill is often employed, but mercurial inunction much more frequently. By the generality of practitioners thefe are ufed in nearly the fame manner as for the venereal difeafe. Opium is by many given with the internal mercurials, but, probably, with great impropriety. We fhould always prefer the ufe of calomel in fuch dofes as to procure occafional difcharges downwards during the whole of the treatment. In reality, mercury has always been found to fucceed beft, when managed fo as to evacuate the inteftines very confiderably during the greater part, if not the whole, of the courfe; and experience has the fanction of reafon in this inftance. No medicine whatever poffeffes the power of mercury in the excitement and extraordinary increafe of all the fecretions, particularly thofe of the bowels. If the fecretions, which certainly take place during the exhibition of mercury, are allowed to accumulate, to ftagnate, and to become acrid in the bowels,

much

much mifchief muft unavoidably prove the refult; a fever, perhaps a dyfentery, but certainly an exacerbation of all the previous fymptoms muft be the confequence. On thefe occafions, the bowels do not, in general, appear to poffefs activity fufficient to evacuate themfelves to a proper degree, and, therefore, require the affiftance of art. In the exhibition of mercury for the venereal difeafe, it is the unanimous opinion of the faculty, that every precaution fhould be taken to prevent the tendency of that medicine to the bowels, as a circumftance very unfavourable to its general effect on all the fecretions of the fyftem, by which it may be fuppofed to operate in the cure of the venereal difeafe. This, no doubt, is extremely juft, but has not the leaft analogy with the ufe of mercury in the liver, as the fecretions of that vifcus and the other bowels are alone concerned in the treatment, at leaft in a primary view; nor does it appear that the excitement of an increafed action in the other organs of the bowels is at all inimical to that of the liver. Even in the foundeft ftate of this vifcus, from whatever fource the bile may immediately flow, very copious difcharges generally follow the exhibition of a purgative; and it may be prefumed, that it does fometimes come from

the

the liver on such occasions. Too much, in
reality, cannot be said against the use of opium
during the mercurial course; nor does the
common, and undoubtedly the best, excuse
for the exhibition of opium ever exist to any
insupportable degree in chronic affections of
the liver. Some degree of pain generally
does affect various parts in these disorders;
but it is never such as requires the alleviation of opium; nor would opium, on most
occasions, afford any relief whatever to these
pains. Another caution is indispensibly necessary here. Instances do occur, in which
the largest quantities of mercurials have been
used without sensible effects of any kind.
Should the mercurials, on such occasions,
have been allowed to (what is called) *pass off*
by the intestines, its failure will be very readily attributed to that circumstance; or to its
combination with a purgative medicine, when
this may have been used. It is certainly an
extraordinary circumstance, that very large
quantities of mercury, exhibited in the various forms of ointment, calomel, and mercurial
pill, should sometimes not only make no im-
pression whatever on the seat of the disorder,
but should fail to affect even the salivary
glands, and every other part, in the slightest
degree. This apparent inertness of the medicine,

dicine, in venereal cafes, has often been attributed to the infirm ftate of the patient; but although, in fuch habits, it may fail in its general effects on the fyftem, its operation on the falivary glands has been favoured by thofe circumftances of debility. It is poffible, though not very probable, that in the liver cafes in queftion, the patients may have deceived the practitioner. For this mode of failure, indeed, there is no remedy. Thefe deceptions, however, are very frequent in Bengal, where the practitioner's attendance is held of lefs value, as of no expence to the patient.

It has been already obferved, that the affection of the mouth, in venereal cafes, is juftly thought to be unneceffarily, if not pernicioufly, retarded, by a combination of laxative medicines; and that although this effect is as certainly retarded by fimilar means, in affections of the liver, yet it is not of equal prejudice in the cure, for the reafons which have been ftated before. We fhould inculcate this, as one of the moft important circumftances in the treatment of the difeafe, and a circumftance that does not always receive the attention which it deferves. Authors have fuppofed that mercury is not adapted to every ftage of thefe complaints;

and

and to thofe, who have not been much converfant in affe
ctions of the liver, the ufe of
mercury, after the fuppofed formation of
matter, would appear an unwarrantable practice, and in direct contradiction to the teftimony of many writers, but particularly of
Doctor Clarke. Were the validity, however,
of this opinion granted, yet the difficulty
would ftill be as great as ever; for how is the
exiftence of abfcefs to be afcertained? We
may venture to fay, not beyond precarious conjecture. The hectic heats and partial fweats,
not to mention the lefs important fymptoms,
which have been received as indicatory of
abfcefs or abforption, afford no criterion for
the regulation of our practice in this refpect;
for thefe may, and generally do, occur in
every fpecies of affected liver, at one period
or other of a diforder of any duration, and
yield as readily to mercury as any other
obvious form of the liver. Hectic heats,
partial fweats, a general decay of embonpoint, colour, and ftrength, are often the very
firft fymptoms of a difeafe in that vifcus.
But, independent of the inutility of fuch a
diftinction, as far as it relates to practice, it
cannot in reality be allowed, that mercury is
ufelefs, much lefs prejudicial, in thofe cafes
where matter has certainly been formed. Affections

fections of that viscus will seldom be found
partial: while one part has advanced to the
stage of abscess, another may have arrived to
within some degrees of it, and a third may be
only in an incipient stage; both the latter,
however, avowedly curable by mercury. The
mischievous effects of this medicine in real
cases of abscess should, one might suppose, be
ascertained, beyond all possibility of doubt,
before a practitioner would venture to pro-
scribe its use, when the earlier stages of the
disorder may still be supposed to exist in
other parts of the liver. It may be asserted
very positively, that experience combats such
prejudices against the use of mercury, and
that they have not even the sanction of any
plausible theory. It has been imagined, that
the absorption of the matter creates all the
general mischief in the habit, which results
from internal abscess; yet it is obvious, that
the matter must form a passage for itself
through some other channel, if it should not
be absorbed. The consequence of this may
prove of either a more or less favourable termination than that by absorption; and it will
be allowed, that the evacuation of the matter,
through the means of absorption, is not necessarily fatal, but will often cure the patient.
In what manner can mercury, under any of

L                    these

thefe circumftances, prove of difservice? If the matter fhould have a tendency to make its way through the biliary ducts into the in- teftines, which is probably the moft favourable courfe, will the action of mercury direct its progrefs through the diaphragm and lungs; or promote its difcharge ftill more deterioufly into the cavity of the abdomen? No theorift will torture reafoning into fuch a conclufion. On the contrary, mercury, from its powerful action on the abforbents, may be reafonably expected to favour the cure of abfcefs through their means, at the very time that it prevents the formation of more matter, or frefh abfceffes, by its general effects on other parts of the liver: and next to the efforts of mere nature, we may believe the only chance of the patient's falvation is to be fought for in the ufe of mercury. It may be faid that this drug will increafe the debility, and tend to render the falutary efforts of nature lefs powerful; but thofe who have had much experience of its ufe in warm climates, are well affured that its effects on no occafion are fo confiderable in this refpect, as have been generally apprehended; that it has even a contrary tendency, if fulnefs and ftrength of pulfe may be allowed to ftand as proofs of it; and that, although there may be one period

riod after the ceffation of the ftimulus, when
a proportional collapfe is fuppofed to take
place, yet, that it is innocuous in its remote
and ultimate confequences, may be prefumed
from the rapid embonpoint and fenfe of
health, which take place fooner or later in
moft inftances. It will refult from thefe
obfervations, that it is not eafy to afcertain
the exact period of the formation of matter
in that vifcus; and that under the utmoft certainty of its exiftence, the curative indications muft ftill continue nearly the fame, unlefs in cafes where the lancet may be fuppofed capable of reaching the feat of the
abfcefs. No doubt, on fuch occafions, a bold
incifion will often refcue the patient from
inevitable death. Mercury, therefore, is applicable to almoft every ftage of every affection of the liver. In chronic cafes, it may
be proper to ufe it only in the *alterative*
courfe, as it is called in Europe, until the
mouth is confiderably affected, refuming it
fome time after, if the fymptoms fhould not
vanifh. It is immaterial what medicine is
combined with the mercury, provided it be
not opium, or any other that may have a
tendency to check the fecretions. Perhaps
antimonials form a favourable combination;
we fhould, however, give the preference to

purga-

purgative medicines, fuch as cathartic extract, refin of jalap, fcammony, &c. Medicines, which are calculated for the relief of
particular fymptoms in other diforders, will
anfwer equally well, when fuch fymptoms
occur in affections of the liver, but need not
be enumerated here. The generality of practitioners allow too much latitude in diet.
Dyfpeptic fymptoms are very ufual attendants
on all affections of the liver, and are fometimes the only obvious appearances. In fuch
inftances it is not difficult to conceive the
probable fatality, or at leaft inutility, of the
medicines generally prefcribed in thofe cafes.
But with refpect to diet, certainly no means
whatever could fave the patient under the
courfe of porter and beef-fteaks, not only
allowed but ftrongly recommended by one of
the oldeft profeffors in the univerfity of Edinburgh. Such a diet would inevitably increafe
every form of difeafe in that vifcus. Surely
there are phyficians who would efteem it very
extraordinary, even in mere dyfpepfia, in Europe. The liver, in every form, requires the
ftricteft confinement to a fpare and vegetable
diet. The drink fhould not contain fermented or fpirituous liquor of any kind; and
this diet fhould be continued for a long time
after the cure of the diforder; in reality,
during

during life, if the patient be fubject to relapfes, as they often are. Exercife in the open
air, as well as a change of air, is of the utmoft
importance in thefe cafes.

Such are the outlines of the treatment in
chronic cafes. The means are nearly the
fame in acute attacks, but ufed with infinitely more vigour, in proportion to the fud-
dennefs and violence of them. The moft
acute forms of liver happen lefs frequently
in Bengal than at Madras; but they do fometimes occur in the former. Sometimes a
fudden fwelling, pretty much circumfcribed
in the region of the liver, is the firft intimation which the patient receives. On all thefe
occafions, mercurial ointment fhould be rubbed into the part, and over moft part of the
furface of the fkin, without the lofs of a
fingle moment, and continued without intermiffion, and without any fcruples refpecting
the quantity, until the mouth be decidedly
affected. If the fymptoms do not remit or
vanifh, when this takes place, it is probable
that an abfcefs, or fome permanent and perhaps fatal affection will enfue. The purgatives are equally neceffary in thefe inftances,
as well as the diet. Farinacea, however, are

L 3                    better

better calculated for thefe forms of the complaint, than the frefh vegetables.

The writer is neceffarily compelled to conclude this article, which contains hints perhaps too general, and certainly too confufed. He cannot expect much reliance on his individual authority in thefe cafes, until the proofs be publifhed; but he will clofe this paper with fome remarks from an authority that was as much refpected as a man and a phyfician in India, during his ufeful life, as any medical man, perhaps, in any part of the world. Doctor Paifly, furgeon-general at Madras, enjoyed fuch an extenfive fame, that he was confulted and even vifited by fick Europeans from every part of India. Medical practice in that country is much indebted to his judgment, fagacity, and original practice, for the moft important improvements. The following is a letter, which he wrote to a young gentleman of the Bengal medical eftablifhment, who has fince rifen to eminence as a practitioner in Calcutta.

Copy

Copy of a Letter from Doctor PAISLY.

" SIR, Fort St. George.
" I HAVE been favoured with yours by
" ——— ———; and as your letter fhews a
" candid inquifitivenefs that merits informa-
" tion, I fhall be fomewhat explicit on the
" fubject. Mr. ———'s cafe is one of thofe
" that occur every day in this country, of-
" tener than is imagined in other hot cli-
" mates, and I believe frequently in camps;
" alfo at fome particular feafons in Europe,
" when bilious diforders prevail; but is ge-
" nerally overlooked, on the fuppofition that
" fluxes are fimple difeafes, arifing from in-
" fectious miafmata, putrid bile, fomething
" acrimonious, or crudities in the prima viæ,
" and of courfe are treated with emetics,
" laxatives, antifeptics, and blunters. In-
" deed, if thefe were only fuch fimple caufes,
" we might generally expect effectual cures
" from a perfeverance in fuch medicines,
" properly adapted. But on the contrary,
" we find them run on to a great length,
" and often a precarious iffue; and to our
L 4 " morti-

" mortification find that such treatment
" proves palliative only for a day. This
" being evidently enough the cafe, we ought
" naturally to conjecture, that the fource of
" their obftinacy muft be looked for beyond
" the inteftinal canal. Such is really the
" fact. A deep-rooted obftruction generally
" fupports the difeafe, in fpite of unwearied
" evacuations; and all affiftance becomes
" only palliative, until that is removed. A
" want of attention to this circumftance
" allows fluxes to run on to their fecond or
" third ftages, as they are called, which in
" reality are no more than different ftages of
" duration and weaknefs: for the difeafe has
" been uniformly the fame from the begin-
" ning, an obftructed liver and mefentery, but
" chiefly the former, with fome degree of
" inflammation; the flux itfelf being no more
" than a fymptom of the difeafe, and an
" effect of difeafed fecretions.

" We have feldom occafion to be doubtful
" of the exiftence of this caufe; for if we
" have not pain to direct us, an experienced
" touch will difcover the obftruction, the
" tendernefs and enlargement of the liver.
" I fay an experienced touch, becaufe it re-
" quires a frequent practical examination to
" fatisfy one refpecting the different degrees
" of

" of hardnefs, firmnefs, and fenfibility, that
" conftitute a difeafe of that bowel; how-
" ever, appearances are feldom fo equivocal
" as to puzzle practice in general. I will
" venture to affirm, that the grand fource of
" health and difeafe is centered in the natu-
" ral or difeafed condition of the liver; and
" that even chronic and lingering illneffes
" arife, in a confiderable degree, from fome
" defect there. In many acute diforders it
" has alfo its fhare. But in every kind of
" ficknefs, whether local or general, that is
" peculiar to this country, it is material to
" examine it; for no perfect cures can be
" made, nor relapfes prevented, without hav-
" ing a ftrict eye to it. The flux he has la-
" boured under for thirteen months might
" originally have been cured in as many
" days; and even now will give but little
" trouble, as he has ftrength enough left to
" bear the operation of medicines.

" This cafe, from the beginning to this
" time, has been an obftructed liver, and
" could at no time have been effectually
" cured, but by mercurial deobftruents:
" though a ftrict diet, exercife, palliative and
" laxative medicines, when the inflammatory
" tendency ceafed, to blunt and evacuate
" bile, difeafed in its fecretion, might have
" given

" given relief for a time, and produced a
" temporary cure. But thefe never could
" reftore him to health, nor that vivid clear-
" nefs of countenance peculiar to it; nor
" even prevent the relapfes he has been fub-
" ject to during fo protracted an illnefs; at
" many different periods of which, had his
" feelings been queftioned, they would have
" been fufficient to have pointed out the
" fource and caufe of his difeafe. I find, by
" his own account, that his cafe in the be-
" ginning was a good deal inflammatory.
" At prefent his liver is only enlarged and
" hard, but no pain there. His flux is ac-
" companied with gripes and tenefmus, and
" his look bloated and fallow; his urine too
" is very high-coloured; this appearance,
" by the bye, never fails to attend liver dif-
" orders of any confequence; though it now
" and then happens, in nervous habits, in
" difeafes of the bile itfelf, or according to
" the fituation of the obftruction, that it is
" either pale, or but little difcoloured; but
" in general it is a material index to difeafes,
" where bile prevails. When obftructions
" of the liver exift, the firft digeftion goes
" on very imperfectly; therefore the firft in-
" dication in his cafe, is to clear the bow-
" els of bile, phlegm, and other caufes of irri-
                                    " tation;

" tation; the fecond, to remove the obftruc-
" tion of his liver; and laftly, to recover the
" ioft tone of his inteftines, and to ftrengthen
" his fyftem in general.

" In recent cafes, caftor oil, where a tenef-
" mus is troublefome, is a good purge, and
" effectually difengages much glutinous bi-
" lious ftuff from the duodenum and the
" colon; and as it enters not the blood, it
" creates little difturbance in the fyftem.
" It is, therefore, I fay, in recent cafes,
" an eligible purge; but in this, where the
" bowels have been much relaxed and weak-
" ened, rhubarb, quickened with calomel or
" foluble tartar, is better to anfwer the firft
" intention, repeated once or oftener, ac-
" cording to the nature and complexion of
" the excretions.

" Proper evacuations having been made,
" the fecond indication is to be anfwered by
" mercury, adminiftered but flowly, as his
" complaints are of a long ftanding, and his
" habit much relaxed. However, one dram
" of mercurial ointment will be rubbed into
" his fide daily, and ten grains of a pill
" compofed of the mucilage mercurial pill
" and ipecacuan, equal parts, will be given
" him night and morning; or the follow-
" ing :

2                                    " ℞ Spec.

" ℞ Spec. aromat. gr. vi. vitr. cerat. antimon.
 " Cal. opt. evigat. āā gr. iii. Conf. al-
 " kerm. q. f. f. bolus m. et v. Sumend.

" Thefe are to be continued till the mer-
" cury fixes in his mouth. By thefe means
" the obftruction will be gradually removed,
" and the load of ferous humours that flow
" to his guts, in their prefent relaxed and
" irritable ftate, will be derived to his
" mouth; by which means the bowels will
" gain a truce from evacuation. As the
" obftructions remove, the urine and ftools
" will grow more natural, and every fecre-
" tion lefs difeafed.

" In thofe chronic obftructed cafes, the
" friction ufed in the application of the oint-
" ment is not without efficacy.

" As you defire my opinion of opiates and
" aftringents, I fhall give it you from expe-
" rience in a few words. Opiates are dan-
" gerous medicines in fluxes, and are always
" to be ufed with the moft fearful caution.
" Early in the difeafe, where the cafe is pu-
" trid or inflammatory, they fhould be ufed
" at no folicitation whatever. Sufpending
" the evacuations for a night may bring on
" a fatal mortification. In more chronic
" cafes, a freedom from gripes, the fmell
" and nature of the evacuations by ftool, are

3                                          " the

" the only criterions to direct the ufe of
" them; for no retentions are to be made
" of any thing putrid or acrimonious. In
" fhort, in a climate where all the capital
" difeafes arife either from putrid bile or ob-
" ftruction, nothing muft be fhut up.

" As for aftringents, they are often dan-
" gerous medicines in fluxes, and always
" precarious, and can be ufed with no kind
" of fafety, until all obftructions are re-
" moved, with every thing that is putrid or
" difeafed. In acute cafes they have fatal
" effects; in the more chronic cafes they
" hamper and protract the diforder, with all
" its concomitants, gripes, tenefmus, &c.
" When the difeafe becomes a fimple di-
" arrhœa, without gripes, or with them
" arifing only from wind, equal parts of
" conf. prun. filv. and the conf. aurant. or
" confect. cardiac. is an elegant and fafe
" aftringent. The tinct. cort. cafcar. cold
" bathing and exercife recover the conftitu-
" tion from a ftate of relaxation.

" Having made thefe remarks on opiates
" and aftringents, I come next to the third
" indication in the cafe of Mr. ———. After
" the obftruction in his liver is removed,
" and other complaints abated, the third
" indication is to be anfwered by gentle
" bitters

" bitters of the leaſt aſtringent kind.  Bark,
" which on moſt occaſions of weakneſs and
" relaxation is an invaluable medicine, in
" liver caſes is a poiſon.  The flor. cham.
" ſem. cam. and ſal polychreſt, each one
" drachm, will make a couple of cups of
" bitter tea, to be drunk forenoon and after-
" noon.  This medicine, with proper exer-
" ciſe and diet, will be ſufficient to re-
" eſtabliſh him.

"  —— ——, whom you may have ſeen,
" laboured under a diſorder of this kind, but
" worſe, being accompanied with much
" weakneſs, and the loſs, in a great mea-
" ſure, of the uſe of his extremities.  He
" is now in perfect health.  The ſource
" and cauſe of his complaints had alſo been
" overlooked.  He had, like Mr. ——, tri-
" fled for many months with the uſual me-
" dicines in fluxes, until he was reduced to
" extreme weakneſs.  Theſe are two inſtan-
" ces of liver fluxes, which you have ſeen.
" I have met, within theſe few years, with
" numberleſs unfortunates from camps, and
" different parts of the country, in the very
" ſame predicament, who have with diffi-
" culty ſurvived this wreck of conſtitution.
" But it is impoſſible to enumerate the va-
" riety of complicated bilious diſorders, that
                                        " practice

" practice daily prefents, with obfervations;
" nor can I in a letter convey a proper idea
" of the nice and peculiar treatment difeafes,
" arifing from putrid bile and obftruction,
" require. Different circumftances and con-
" ftitutions diverfify; but all capital difeafes
" require the moft circumfpect attention,
" and a confcientious attendance on them.
" Omiffions and miftakes are not to be re-
" medied by any future effort; efpecially
" where putrid bile lurks in the habit.
" Difturbing it fuddenly by evacuations, or
" putting it in motion before it is well di-
" luted and corrected, often proves, like poi-
" fon, fuddenly fatal. Of this I have feen
" inftances in men, who were not at the
" time bed-ridden. As a ftriking inftance
" alfo of the general attention that is necef-
" fary in the moft trivial complaints arifing
" from bile, I can affure you I have known
" what are trifling medicines prefcribed in
" ailments which were confidered as equally
" trifling, productive of very ferious difor-
" ders. For example: I have feen a courfe
" of bitters, ordered in what was imagined a
" windy relaxed ftomach, produce liver ob-
" ftructions, and, in inflammatory habits, a
" real inflammation of that bowel. Such
" circumftances awaken a man's caution;
" but

" but it is a practical fact, that no strong or
" aftringent bitter can be used here with im-
" punity, unlefs the cafe is fimply relaxation.
" In mixed diforders, the gentleft of the kind,
" with neutral falts, are only fafe. Indeed it
" is ftill, perhaps, a problem in phyfic, how
" long the ufe of aftringents may be conti-
" nued with fafety, or without danger of en-
" tailing fome additional diforder. Before I
" conclude this letter, though fomewhat fo-
" reign to the fubject of yours, I cannot avoid
" putting you on your guard againft a difor-
" der of the liver, which, from its being over-
" looked, I have once feen in Europe, and
" feveral times here, attended with fatal con-
" fequences. The diforder I mean, is what
" may be termed a liver cough. The ob-
" ftruction, in this cafe, is pretty generally at-
" tended with inflammation and pain, though
" feldom acute, unlefs preffed with the fin-
" gers, or when the external membrane is
" alfo affected; but it oftener happens with-
" out pain or inflammation. The cough,
" though only a fymptomatic complaint, is
" the *misleading symptom* of the difeafe. The
" patient pronounces his own cafe a cold, and
" is put on a courfe of ineffectual pectorals,
" takes exercife, and fhifts his fituation for
" health, until his liver either fuppurates, or
" becomes

" becomes an indolent mafs of irrecoverable
" obftruftions. In very irritable inflammatory
" habits, any miftake at the commencement
" of the difeafe is of the moft dangerous con-
" fequence. The liver, the diaphragm, and
" the lungs adhere and fuppurate, and a pu-
" rulent fpitting fucceeds; though, inftead of
" a fmooth uniform pus, the fubftance of the
" liver is expectorated by a deep hollow
" cough, in form of glandular membraneous
" appearances, mixed with purplifh difco-
" loured blood, of a parenchymatous look.

" This diforder, like all other inflamma-
" tory diforders of the liver, is very tractable
" in the beginning, by evacuations, relax-
" ing antiphlogiftic medicines, and mercury.
" Such cafes as the above I have met with,
" and have been happy enough to effect fome
" cures, even in that advanced ftage.

" It is to be obferved, that in all confirmed
" difeafes of the lungs of any ftanding, the
" liver is always affected; but in this difor-
" der the lungs are only the fecondary object,
" and never give any trouble, if the obftruc-
" tion of the liver be removed, as in them
" there are neither tubercles nor infarctions.
" The breathing, except in inflammatory cafes,
" is never affected; and the fymptomatic
" complaints, cough and pain in the fhoul-

M " der,

" der, may always be mitigated by lying on
" the back, with the head low and the legs
" raifed.

" To the above I fhall annex another dif-
" order, which is entirely of the liver kind.

" Agues are not frequent in the dry fandy
" foil of this coaft; but quotidian remit-
" tents often happen here from inflammatory
" obftructions of the liver; I fay remittents,
" becaufe the fever never goes entirely off,
" though the quotidian ague fits are very
" regular in their attacks. The patient in
" fuch cafes tells you only that he has got a
" fever and ague, and a pain at the pit of the
" ftomach; but his report is not to be trufted,
" without further examination of him, both
" in a lying and ftanding pofture. On pref-
" fing with your fingers, from the ftomach
" to the right fide, he will complain of much
" pain and tendernefs, and his urine will be
" very high-coloured; fometimes his right
" fhoulder is affected, and generally he has
" a liver cough. Here is a complicated
" cafe, though there are no contra-indica-
" tions in the cure of it. The inflammation
" and obftruction of the liver are the proxi-
" mate caufes of the diforder, and its dif-
" ferent fymptoms. Venefection is imme-
" diately neceffary, and, if the ftomach feems
" loaded,

" loaded, an emetic may be fafely given after
" it; but it fhould be fuch as will operate
" eafily, and open, the belly. Such I have
" found the following in all bilious com-
" plaints:
 " ℞ Vin. antim. v. ipecacuan. āā ℥ſs.
 " Oximel. fcil. ʒvi. M.
 " Notwithſtanding the continual fever,
" ʒi. of mercurial ointment muſt be rubbed
" into the fide morning and evening, after
" fomenting it. Saline draughts, with nitre
" and tartar emetic, every two or three hours,
" and bleeding repeated if neceſſary. The
" ague and cough require no attention; theſe
" will diminiſh as the mercury takes effect,
" and entirely ceafe when the fpitting comes
" on. Any bilious attacks that may happen
" during this interval are to be relieved by
" emetics, caſtor-oil or foluble tartar, and
" glyſters, as circumſtances may require.
 " We meet here with quotidians and ter-
" tians, ariſing from obſtructed livers, where
" bark is never neceſſary, but to prevent re-
" lapfes. Some time fince I met with a
" quartan of two years ſtanding, attended
" with a very irregular ſtate of the bowels
" on every acceſſion of the fit, with fuch an
" aſtoniſhing fecretion of bile as to produce
" a fevere cholera morbus. After reducing
 " the

" the liver obstructions by mercury, and the
" patient in a proper train for astringents,
" his cure was effectually compleated with

 " ℞ Conserv. aurant. ℨi.

  " Pulv. cort. peruv. ℥ſs.

  " Serpent. virg. ʒii.

  " Tart. vitr. ʒi.

  " Syr. aurantior. q. ſ. ut f. elect.

" Of this he took half an ounce in the
" day, and continued it for ſome time; and
" every third night, when his excretions
" were not free, he took ten grains of a maſs
" of pills, compoſed of equal parts of ſoap,
" aloes, and calomel.

 " On the ſubject of agues, I ſhall obſerve
" to you, in general, that bark is ſeldom ne-
" ceſſary in the cure of them, and often un-
" ſafe, except in caſes of great weakneſs, re-
" laxation, or where the nervous ſyſtem is
" much affected : on the contrary, evacuants,
" neutral ſalts, and deobſtruents, are ſeldom
" ineffectual. However, where the indica-
" tion is not ſtrongly marked, and when dif-
" ficulties ariſe about the propriety of exhi-
" biting particular medicines, the following
" criterion, with reſtrictions, may ſerve as a
" guide.

 " When the bark does not ſucceed in
" ſtopping the fits and preventing relapſes,
         " deob-

" deobftruents, neutral relaxing medicines,
" with evacuations according to circumftan-
" ces, are more likely to anfwer. On the
" other hand, if they fail, bark, bracing me-
" dicines, and nervous, will become more
" neceffary; but whether in agues, or in dif-
" eafes induced by obftructions of the vif-
" cera, if the urine is high-coloured in the
" interval of the fit, or during the courfe of
" the difeafe, aftringents of every kind are to
" be avoided; for infinite mifchief may be
" done by them, efpecially if the difeafe be
" inflammatory, or if putrid bile lurks in the
" primæ viæ, from which alone wonderful
" and mifleading effects on the fyftem are
" often produced. Sudden fevers are lighted
" up, choleras are induced, convulfions,
" fpafms, and all the variety of nervous
" affections, according to the degree of its
" acrimony, and the particular idiofyncrafy
" of the patient.

" The cafe of a gentleman, who was un-
" der my care not many days fince, will ex-
" emplify this. A healthy florid young man
" was taken fuddenly with fever. A dry
" fkin, his tongue as white as a fheet, his
" urine like porter, and of a ftrong alkaline
" fmell; great heat and fulnefs about the
" præcordia; his pulfe fometimes high, fome-

M 3                        " times

" times low, but always quick; great and
" reftlefs agitations in his whole frame;
" fometimes the deepeft dejections of fpi-
" rits, even to crying; fometimes ravings,
" horrors, and general fpafms; fometimes
" calling for wine to fupport him under thefe
" depreflions, at other times for water to
" allay an unquenchable thirft. All thefe
" fymptoms, and the tranfition of them, were
" difplayed during the firft vifit I made him,
" which happened a few hours after his
" being feized. This appeared clearly enough
" to be a cafe of putrid bile, operating on
" the fyftem, and is one of thofe cafes,
" where the patient is either out of danger
" or extinguifhed in a few hours. The whole
" indications of cure, in fuch cafes, are to
" dilute and evacuate the irritating caufe,
" and by tempering medicines to quiet the
" difturbances raifed. Stimulating medicines
" have no place here, notwithftanding the
" variety of nervous affections; they unavoid-
" ably would aggravate every fymptom.
" Emetics ruffle too much to be ventured
" on, when the whole fyftem is diftracted,
" befides the danger of their fetting in mo-
" tion, at once, a deluge of putrid bile, which
" is never without inftant rifk. For this
" gentleman I ordered a purging glyfter im-
                              " mediately;

3

" mediately; and his legs, &c. to be fomented
" with falt water, until a folution ℥i. of
" manna, and ℥ís. of foluble tartar, in ℥x.
" of water could be fent him. Of this
" he was directed to take four fpoonfuls
" every hour, until it operated three times,
" and brought away fome frothy bile, of an
" intolerable ftench, and like the workings
" of a beer cafk. This was far from fuffi-
" cient to produce very confiderable effects;
" however, he was fenfibly relieved, and it
" was no fmall fatisfaction, from the nature
" of the excretions, to find the difeafe in my
" power. As foon as the effects of the laxa-
" tive were over, I directed him to take,
" every three hours, four fpoonfuls of the
" following

" ℞ Camphor. nativ. gr. x.
" Amygdal. dulc. decort. dr. ii.
" Mucilag. gum. arab. falin. ℥x.
" Pulv. nitr. ʒ ís.
" Tart. emetic. gr. i.
" Syr. facch. ʒ ii. M.

" This to be continued in the interval of
" purging; and to ufe clear rhenifh whey as
" common drink ad libitum. He paffed a
" more tolerable night, and lefs oppreffed,
" lefs reftlefs, and his head lefs affected. The
" day following he was ordered another more

" active

" active purge, compofed of fena ʒi. manna
" ʒiii. foluble tartar ʒiii. fem. camin. ʒfs,
" to be infufed in ʒviii. of boiling water;
" one half to be taken early in the morning,
" the other two hours after. This operated
" five times, the ftools of the fame nature
" and fmell, but plentiful. Every fymptom
" abated confiderably after the operation of
" the purge; his urine and tongue remained
" the fame, but he was much compofed;
" few fpafms, few attacks of depreffion of
" fpirits, his fkin moifter, and his pulfe regu-
" lar and lefs frequent. It only remained
" now to prepare the remainder of the bile
" for expulfion; he was therefore directed to
" drink plentifully, to take his medicines
" regularly for a couple of days, and the
" third day in the morning the fame purge
" was repeated, with the addition of ʒfs. of
" fena. It operated very effectually, and
" brought away much glutinous ftuff and
" bile, and the evening of that day his urine
" became natural and clear, his tongue but
" little changed, and every complaint va-
" nifhed. Nothing further was neceffary,
" but to leave a purge with him, to direct
" him to continue the rhenifh whey, and ufe
" light diet until he recovered fome ftrength.
" But as this diforder, with the mort de
                              " chiens,

9

" chiens, and many others of the putrid bi-
" lious kind, originate with the liver, relapfes
" cannot be prevented without removing
" every obftruction, and reftoring the fecre-
" tions. For, although all acrimonies and
" putrid accumulations arifing from difeafed
" fecretions may be removed by well-timed
" evacuations, yet frefh collections and indi-
" geftions may foon occafion the fame fcene
" to be acted, if obftruction is not removed,
" and healthy bile reftored, to perform na-
" ture's firft and greateft operation in the ani-
" mal œconomy. We fhould therefore never
" think it fufficient to fave the patient from
" immediate and preffing danger. The latent
" defect, the original of all, it is alfo incum-
" bent on us to remove. The ftate of the
" ftomach, the vifceral glands, fecretions, and
" urine, are to be nicely attended to, and, as
" circumftances may require, gentle mercu-
" rial deobftruents, gentle bitters, with neutral
" falts, are to be adminiftered, and occafional
" evacuations, with a ftrict diet, to be recom-
" mended, until the conftitution and health
" are perfectly reftored; otherwife our fervices,
" rather flattering than ufeful, if I may be
" allowed to ufe the comparifon, are like the
" fplendid actions of the ftatefman or the gene-
" ral, which acquire them reputation and eclat,
" but

" but are productive of no folid or lafting
" advantage in the fyftem.

" I could furnifh you with a variety of
" fuch cafes, and with a multitude of hifto-
" ries of other complicated bilious diforders,
" which come daily under my notice here;
" but they would fwell this letter to a vo-
" lume, which has already, indeed, exceeded
" much the bounds I intended; I fhall there-
" fore only further obferve, in general, that
" bile in different ftates and fituations, and in
" different conftitutions, puts on the appear-
" ance of, and apes, almoft every diforder;
" and of confequence much experience and
" attention are requifite to difcriminate pro-
" perly its effects and operation in this coun-
" try, where there are but few fimple fluxes,
" fimple agues, or fimple coughs, or, indeed,
" but few fimple difeafes; and you may be
" affured, that where the liver and primæ
" viæ are not confidered as the grand fources
" of difeafes, continual and ferious blunders
" will be committed in practice.

" Mercury, in judicious hands, is a fafe
" and tractable medicine, and as it is the only
" fafe and powerful deobftruent in glandular
" obftructions, it is of confequence the only
" medicine to be depended on in thofe latent
" defects of the fyftem, which entail difeafes

" or

" or impede recovery; however, it often re-
" quires affiftance from other medicines, from
" exercife, from fpas, or from medicated
" aqueous medicines, which wafh the glands
" and increafe the fecretions; or, in general,
" it requires affiftance confiftent with the
" effects difeafes have had on the conftitu-
" tion. For example; in venereal and other
" habits, where the folids are much relaxed,
" the blood poor and flimfy, mercury, with-
" out bark, will fpread every ulcer, induce
" febricule, and aggravate the fymptoms. In
" acute difeafes, evacuations, neutral and re-
" laxing medicines, render it fafe and effec-
" tual. In irritable habits it requires ma-
" nagement, as its operation is chiefly on the
" folids; but the idea of its injuring the con-
" ftitution, or diffolving the crafis of the
" blood, is without foundation : on the con-
" trary, it is too apt to leave behind an in-
" flammatory diathefis."

*Note.*—Part of the copy is illegible here.

" You may obferve, I have been fomewhat
" general in my anfwer to your's, but I was
" willing to throw into it as much informa-
" tion as the compafs of a letter would ad-
                                        " mit.

" mit. Upon the whole, it contains a few
" hints, which you may in future improve to
" your advantage; though to have been cir-.
" cumſtantially minute was impoſſible."

*Note.*—The copy is again illegible here.

F I N I S.

www.ingramcontent.com/pod-product-compliance
Lightning Source LLC
Chambersburg PA
CBHW021803190326
41518CB00007B/433